Reviews and critical articles covering the entire field of normal anatomy (cytology, histology, cyto- and histochemistry, electron microscopy, macroscopy, experimental morphology and embryology and comparative anatomy) are published in Advances in Anatomy, Embryology and Cell Biology. Papers dealing with anthropology and clinical morphology that aim to encourage cooperation between anatomy and related disciplines will also be accepted. Papers are normally commissioned. Original papers and communications may be submitted and will be considered for publication provided they meet the requirements of a review article and thus fit into the scope of "Advances". English language is preferred.

It is a fundamental condition that submitted manuscripts have not been and will not simultaneously be submitted or published elsewhere. With the acceptance of a manuscript for publication, the publisher acquires full and exclusive copyright for all languages and countries.

Twenty-five copies of each paper are supplied free of charge.

Manuscripts should be addressed to

Prof. Dr. F. BECK, Howard Florey Institute, University of Melbourne, Parkville, 3000 Melbourne, Victoria, Australia
e-mail: fb22@le.ac.uk

Prof. Dr. F. CLASCÁ, Department of Anatomy, Histology and Neurobiology,
Universidad Autónoma de Madrid, Ave. Arzobispo Morcillo s/n, 28029 Madrid, Spain
e-mail: francisco.clasca@uam.es

Prof. Dr. M. FROTSCHER, Institut für Anatomie und Zellbiologie, Abteilung für Neuroanatomie,
Albert-Ludwigs-Universität Freiburg, Albertstr. 17, 79001 Freiburg, Germany
e-mail: michael.frotscher@anat.uni-freiburg.de

Prof. Dr. D.E. HAINES, Ph.D., Department of Anatomy, The University of Mississippi Med. Ctr.,
2500 North State Street, Jackson, MS 39216–4505, USA
e-mail: dhaines@anatomy.umsmed.edu

Prof. Dr. N. HIROKAWA, Department of Cell Biology and Anatomy, University of Tokyo,
Hongo 7–3–1, 113-0033 Tokyo, Japan
e-mail: hirokawa@m.u-tokyo.ac.jp

Dr. Z. KMIEC, Department of Histology and Immunology, Medical University of Gdansk,
Debinki 1, 80-211 Gdansk, Poland
e-mail: zkmiec@amg.gda.pl

Prof. Dr. H.-W. KORF, Zentrum der Morphologie, Universität Frankfurt,
Theodor-Stern Kai 7, 60595 Frankfurt/Main, Germany
e-mail: korf@em.uni-frankfurt.de

Prof. Dr. E. MARANI, Department Biomedical Signal and Systems, University Twente,
P.O. Box 217, 7500 AE Enschede, The Netherlands
e-mail: e.marani@utwente.nl

Prof. Dr. R. PUTZ, Anatomische Anstalt der Universität München,
Lehrstuhl Anatomie I, Pettenkoferstr. 11, 80336 München, Germany
e-mail: reinhard.putz@med.uni-muenchen.de

Prof. Dr. Dr. h.c. Y. SANO, Department of Anatomy, Kyoto Prefectural Un
Kawaramachi-Hirokoji, 602 Kyoto, Japan

Prof. Dr. Dr. h.c. T.H. SCHIEBLER, Anatomisches Institut der Universität,
Koellikerstraße 6, 97070 Würzburg, Germany

Prof. Dr. J.-P. TIMMERMANS, Department of Veterinary Sciences, University of Antwerpen,
Groenenborgerlaan 171, 2020 Antwerpen, Belgium
e-mail: jean-pierre.timmermans@ua.ac.be

199
Advances in Anatomy, Embryology and Cell Biology

Editors

F. Beck, Melbourne · F. Clascá, Madrid
M. Frotscher, Freiburg · D.E. Haines, Jackson
N. Hirokawa, Tokyo · Z. Kmiec, Gdansk
H.-W. Korf, Frankfurt · E. Marani, Enschede
R. Putz, München · Y. Sano, Kyoto
T.H. Schiebler, Würzburg
J.-P. Timmermans, Antwerpen

Tjitske Heida, Enrico Marani,
and Kamen G. Usunoff

The Subthalamic Nucleus Part II: Modelling and Simulation of Activity

With 54 Figures

Tjitske Heida
Enrico Marani

Department of Biomedical Signals and Systems,
University of Twente,
7500 AE Enschede
The Netherlands

e-mail: *t.heida@el.utwente.nl*
e-mail: *e.marani@utwente.nl*

Kamen G. Usunoff

Department of Anatomy & Histology,
Medical University Sofia,
1431 Sofia
Bulgaria

e-mail: *uzunoff@medfac.acad.bg*

ISSN 0301-5556
ISBN 978-3-540-79461-5   e-ISBN 978-3-540-79462-2

Library of Congress Control Number: 2008927199

© 2008 Springer-Verlag Berlin Heidelberg

This work is subject to copyright. All rights are reserved, whether the whole or part of the material is concerned, specifically the rights of translation, reprinting, reuse of illustrations, recitation, broadcasting reproduction on microfilm or in any other way, and storage in data banks. Duplication of this publication or parts thereof is permitted only under the provisions of the German Copyright Law of September 9, 1965, in its current version, and permission for use must always be obtained from Springer-Verlag. Violations are liable to prosecution under the German Copyright Law.

The use of general descriptive names, registered names, trademarks, etc. in this publication does not imply, even in the absence of a specific statement, that such names are exempt form the relevant protecttive laws and regulations and therefore free for general use.
Product liability: The publisher cannot guarantee the accuracy of any information about dosage and application contained in this book. In every individual case the user must check such information by consulting the relevant literature.

Printed on acid-free paper

9 8 7 6 5 4 3 2 1

springer.com

List of Contents

1	Introduction..	1
2	**The Basal Ganglia**	1
2.1	Pathways Within the Basal Ganglia......................................	2
2.1.1	Direct Pathway..	2
2.1.2	Indirect Pathway	2
2.1.3	Hyperdirect Pathway...	4
2.1.4	Role of the Direct, Indirect, and Hyperdirect Pathways....................	4
2.1.5	Role of Dopamine in the Direct and Indirect Pathways....................	6
2.1.6	Conduction Times of Pathways	6
2.2	Parkinson's Disease...	6
2.2.1	Direct and Indirect Pathways in PD	7
2.2.2	Changes in Neuronal Firing Rate in PD	8
2.2.3	Changes in Neuronal Firing Pattern in PD..............................	9
2.3	Deep Brain Stimulation...	9
2.3.1	Which Neuronal Elements Are Influenced by DBS?.......................	11
2.3.2	Mechanisms of DBS: Hypotheses	11
3	**STN Activity Recorded in Vitro: Brain Slices**	14
3.1	Spontaneous Activity...	15
3.1.1	Single-Spike Mode...	15
3.1.2	Burst-Firing Mode...	17
3.2	Depolarizing and Hyperpolarizing Inputs...............................	19
3.2.1	Plateau Potential...	21
3.2.2	Low-Threshold Spike...	22
3.3	Ionic Mechanisms of a Plateau Potential	23
3.4	Synaptic Inputs..	25
3.5	High-Frequency Stimulation of STN Cells..............................	26
3.6	Intrinsic Versus Extrinsic Properties: Bursts............................	27
3.6.1	Definition of Bursts..	27
3.6.2	Burst Detection Algorithms	28
3.6.3	Network Bursts Using Burst and Phase Profiles	30
4	**STN Activity Recorded in Vitro: Dissociated Cell Cultures**	31
4.1	Experimental Set-up	31
4.1.1	Cell Culture...	33
4.1.2	Measurement Set-up	33
4.2	Spontaneous Activity...	33

| 4.3 | Addition of Acetylcholine | 35 |
| 4.4 | Electrical Stimulation | 37 |

5	**STN Cell Models and Simulation of Neuronal Networks**	**40**
5.1	Otsuka's Model	40
5.1.1	Membrane Dynamics	41
5.1.2	Spontaneous Activity	43
5.1.3	Plateau Potential Generation	45
5.2	Terman and Rubin's Model	52
5.2.1	Membrane Dynamics	52
5.2.2	Spontaneous Activity	53
5.2.3	Rebound Bursts	56
5.3	Comparison of the Otsuka Model with the Terman/Rubin Model	58
5.4	The Multi-compartment STN Model of Gillies and Willshaw	63
5.4.1	Membrane Dynamics	63
5.4.2	Activity Patterns	64
5.5	Intra-nuclear Network Models	66
5.6	Inter-nuclear Network Models	67
5.6.1	GPe-STN Network	67
5.6.2	GPe-STN-GPi-Thalamus Network	71

6	**Comparison of Part I and Part II**	**75**
6.1	Recurrent STN Axons	75
6.2	Inter-neurons in the STN	75
6.3	Fibre Tracts around and in the STN	75
6.4	Ca^{2+} Receptors	76
6.5	Three-Dimensional Modelling	76
6.6	Types of Projection Neurons	76
6.7	Neurotransmitter Input Versus Receptors in the STN	77
6.8	The Pedunculopontine Nucleus	77
6.9	Nigro-subthalamic Connections	77
6.10	Another Cortico-subthalamic Loop	78
6.11	Nissl-Based Subdivision of the STN	78

| **Appendix 1** | Model Parameter Values Otsuka et al. 2004 | 78 |
| **Appendix 2** | Model Parameter Values Terman et al. 2002; Rubin and Terman (2004) | 79 |

| **References** | | **81** |
| **Index** | | **87** |

Abstract

Part I of *The Subthalamic Nucleus* (volume 198) (STN) accentuates the gap between experimental animal and human information concerning subthalamic development, cytology, topography and connections. The light and electron microscopical cytology focuses on the open nucleus concept and the neuronal types present in the STN. The cytochemistry encompasses enzymes, NO, glial fibrillary acidic protein (GFAP), calcium binding proteins, and receptors (dopamine, cannabinoid, opioid, glutamate, γ-aminobutyric acid (GABA), serotonin, cholinergic, and calcium channels). The ontogeny of the subthalamic cell cord is also reviewed. The topography concerns the rat, cat, baboon and human STN. The descriptions of the connections are also given from a historical point of view. Recent tracer studies on the rat nigro-subthalamic connection revealed contralateral projections. This monograph (Part II of the two volumes) on the subthalamic nucleus (STN) starts with a systemic model of the basal ganglia to evaluate the position of the STN in the direct, indirect and hyperdirect pathways. A summary of in vitro studies is given, describing STN spontaneous activity as well as responses to depolarizing and hyperpolarizing inputs and high-frequency stimulation. STN bursting activity and the underlying ionic mechanisms are investigated. Deep brain stimulation used for symptomatic treatment of Parkinson's disease is discussed in terms of the elements that are influenced and its hypothesized mechanisms. This part of the monograph explores the pedunculopontine–subthalamic connections and summarizes attempts to mimic neurotransmitter actions of the pedunculopontine nucleus in cell cultures and high-frequency stimulation on cultured dissociated rat subthalamic neurons. STN cell models – single- and multi-compartment models and system-level models are discussed in relation to subthalamic function and dysfunction. Parts I and II are compared.

Abbreviations

A	Fields of Sano
A	Adenosine receptor
A8,A9	Catecholaminergic areas
ABC	Avidin-biotin-HRP complex
Alent	Ansa lenticularis
AMPA	α-Amino-3-hydroxy-5-methyl-4-isoxazole-proprionic acid
Am(g)	Amygdala
Apt	Anterior pretectal nucleus: dorsal (AD), medial (AM), and ventral (AV) parts
APV	D-2-Amino-5-phosphono-valerate
AWSR	Array-wide spiking rate
AV	Anterior thalamic nucleus
BAPTA	1,2-bis(2-Aminophenoxy)-ethane-N,N,N',N'-tetraacetic acid
bc	Brachium conjunctivum
bci	Brachium of the colliculus inferior
BDA	Biotinylated dextran amine
BG	Basal ganglia
BI	Burst index
BIP	Burst intensity product
bp	Brachium pontis
CaBP	Calcium binding proteins
CB	Cannabinoid receptor
CB	Calbindin
CC	Corpus callosum
cd	Nucleus caudatus
Ce	Capsula interna
ChII	Chiasma opticum
CG	Central grey
Ci	Capsula interna
ci	Capsula interna
Cl	Corpus Luysii
cl	Contralateral

cla	Claustrum
Cm	Corpus mamillare
CM	Centre median
Cml	Ganglion laterale corp. mamillare
Cmm	Ganglion mediale corp. mamillare
Coa	Commissural anterior
Coha	Commissura hypothalamica anterior
Cop	Commissura posterior
Cospm	Commissura supramamillaris
cp	Pedunculus cerebri
CR	Calretinin
Cu	Cuneiform nucleus
Csth	Corpus subthalamicum
ctb	Central tegmental tract of von Bechterew
ctt	Central tegmental tract
δ	Opioid receptor
d	Vesicle containing dendrites
D	Dopamine receptor
DA	Dopamine
Dbc	Decussation of brachium conjunctivum
DBS	Deep brain stimulation
dcv	Dense core vesicle terminals
DIV	Days in vitro
Dlx1/2	Homeobox gene
DNQX	6,7-Dinitroquinoxaline-2,3-dione
E	Embryonic day
EP	Nucleus entopeduncularis
F1	Flat type 1 (boutons)
F2	Flat type 2 (boutons)
Fhy	Fasciculus hypophyseos
Fmp	Fasciculus mamillaris princeps
Fo	Fornix
Fsp	Fasciculus subthalamico-peduncularis
fp	Fibrae perforantes
frtf	Fasciculus retroflexus Meynerti
Fu	Fasciculus uncinatus
GABA	γ-Aminobutyric acid
GAD	Glutamic acid decarboxylase
GAT	Specific high-affinity GABA uptake protein
GC	Gyrus cinguli

GCA	Gyrus centralis anterior
GCP	Gyrus centralis posterior
Gem	Ganglion ectomamillare
GF	Gyrus fusiformis
GH	Gyrus hippocampi
Ghb	Ganglion habenulae
gl	Corpus geniculatum
Glp	Glandula pinealis
glp	Globus pallidus
Glu	Ionotropic glutamate receptor
GP	Globus pallidus
GPe	Globus pallidus externus
GPi	Globus pallidus internus
H,h	H (Haubenfelder) fields of Forel
5HT	5-Hydroxytryptamine
HRP	Horseradish peroxidase
HVA	High voltage activated currents
I	Insula Reilii
i	Nucleus internus gangl. med. corp. mamillaris
il	Ipsilateral
Ins	Insula
ISI	Interspike interval
κ	Opioid receptor κ
Kv3	Type delayed rectifier
L	Calcium channel type
ll	Lemniscus lateralis
Lm	Lemniscus medialis
Lmi	Lamina medullaris interna
Lmm	Lamina medullaris medialis
Lml	Lamina medullaris lateralis
Lp	Posterior limitans thalamic nucleus
LPc	Gyrus paracentralis
LPi	Lobulus parietalis inferior
LR1	Large round type 1 (bouton)
LR2	Large round type 2 (bouton)
LTS	Low-threshold spike
μ	Opioid receptor μ
M,m	Cholinergic receptor
MEA	Midbrain extrapyramidal area

MEA	Multi-electrode array
mGlu	Metabotropic glutamate receptor
ml	Medial lemniscus
mlf	Fasciculus longitudinalis medialis
MPTP	1-Methyl-4-phenyl-1,2,3,6 tetrahydropyridine
mV	Motor nucleus of the nervus trigeminus
N	Calcium channel type
N	Substantia nigra
Nam	Nucleus amygdaliformis
Nans	Nucleus ansae lenticularis Meynerti
Narc	Nucleus arcuatus thalami
Nc	Nucleus caudatus
Nci	Nuclei of the colliculus inferior
NcM	Nucleus commissurae Meynerti
Ndd	Nuclei dorsales disseminati thalami
Neop	Nucleus of Darkschewitsch
NGF	Nerve growth factor
Ni	Substantia nigra
Nic	Substantia nigra pars compacta
Nir	Substantia nigra pars reticulata
Nkx-2.1	Homeobox gene
Nl	Nucleus centralis thalami
Nld	Nucleus lateralis dorsalis thalami
Nlv	Nucleus lateralis ventralis thalami
Nlve	Nucleus lateralis ventralis ext. thalami
Nlvi	Nucleus lateralis ventralis int. thalami
Nm	Nucleus medialis thalami
Nmi	Nucleus mamilloinfundibularis
NMDA	N-Methyl-d-aspartate
NO	Nitric oxide
NOS	Nitric oxide synthase
NP	Pontine nuclei
Nso	Nucleus supraopticus
NR	Subtypes NMDA receptor
Ntg	Nucleus ruber tegmenti
Ntgd	Nucleus ruber tegmenti pars dorsalis
NIII	Nucleus oculomotorius
NVme	Mesencephalicus trigeminal nucleus
6-OH-DA	6 Hydroxy dopamine
ot	Tractus opticus

ω-CgTX	ω-Conotoxin
ω-AgTX	ω-Agatoxin
P	Postnatal day
P	Calcium channel type
pale	Globus pallidus externus
pali	Globus pallidus internus
parahip	Parahippocampal gyrus
PBP	Parabrachial pigmented nucleus
pc	Pedunculus cerebri
PD	Parkinson's disease
Ped	Pedunculus cerebri
Pl	Nucleus paralemniscalis
Pp	Pes pedunculus
PPN	Nucleus tegmenti pedunculopontinus
ppci	Capsula interna pars peduncularis
Pu	Putamen
Pul	Pulvinar
put	Putamen
PV	Parvalbumin
Q	Calcium channel type
R	Calcium channel type
R	Nucleus ruber
RE	Thalamo-reticular cells
RT	Nucleus reticularis thalami
Ru	Nucleus ruber
SC	Colliculus superior
SEM	Scanning electron microscopy
Sg	Suprageniculate nucleus
Shh	Sonic hedgehog
Smg	Gyrus supramarginalis
SN	Substantia nigra
SNc	Substantia nigra pars compacta
SNl	Substantia nigra pars lateralis
SNr	Substantia nigra pars compacta
Sns	Substantia nigra Soemmeringi
Spa	Substantia perforata anterior
SR	Small round boutons
St	Stria cornea
st	Spinothalamic tract
Stri	Stratum intermedium pedunculi

Strz	Stratum zonale thalami
STN	Subthalamic nucleus
T	Calcium channel type
t	Türck's part of cerebral peduncle
T1–3	Temporal gyri
TII	Tractus opticus
Tbc	Tuber cinereum
TC	Thalamo-cortical cells
TcTT	Tractus corticotegmentothalamicus Rinviki
TEA	Tetraethylammonium chloride
Tgpp	Nucleus tegmenti pedunculopontinus
Tpt	Tractus peduncularis transversus
Tri	Trigonum intercrurale
Tt	Taenia thalami
TTX	Tetrodotoxin
Un	Uncus
Va	Fasciculus mamillothalamicus
VA	Ventral anterior thalamic nucleus
VE	Nuclei ventralis thalami
Vim	Nucleus ventralis intermedius thalami
VM	Ventral medial thalamic nucleus
Voa	Nucleus ventro-oralis anterior
Vop	Nucleus ventro-oralis posterior
VPI	Nucleus ventralis posterior inferior thalami
VPL	Nucleus ventralis posterior lateralis thalami
VPM	Nucleus ventralis posterior medialis thalami
VTA	Ventral tegmental area
VIII	Ventriculus tertius
Wnt-3	Homeobox gene
Zi	Zona incerta
II	Optic tract
3D	Three dimensional
IV	Nervus trochlearis

1
Introduction

In order to investigate the dynamics of single subthalamic nucleus (STN) neurons and the activity patterns when STN neurons are connected with other types of (basal ganglia) neurons, in vitro brain slices and networks of dissociated neurons of a specific type cultured on a multi-electrode array (MEA) may be used to identify the ion channels that determine membrane dynamics and to investigate network activity in open-loop configurations. In addition, mathematical modelling provides a powerful tool to test neuron and network behaviour under different conditions. Especially the emergence of deep brain stimulation (DBS) as an effective treatment for the symptoms of Parkinson's disease (PD) has recently motivated the development of several computational models to probe (part of) the underlying network activity as well as the mechanisms of stimulus-induced suppression of PD-induced activity patterns.

Single-compartment, conductance-based models may be developed on the basis of patch clamp studies in which all types of ion channels within the membrane and their dynamics are retrieved. Networks consisting of a large number of different types of neurons and different types of inter-connections may be created using low-dimensional models based on these empirical models.

The network of the basal ganglia and the role of the STN within this network is described in Sect. 2 (this volume). Since a large number of experimental as well as modelling studies focus on PD, changes in network activity under dopamine-depleted conditions are described. An overview of in vitro experiments is given in Sect. 3 and 4 (this volume) in which the different modes of activity of STN cells that were observed are summarized. In Sect. 5, various current STN neuron and network models are summarized in view of the occurrence of bursting activity and synchronization within the network of the basal ganglia, as occurs in PD, and the use of STN-DBS (deep brain stimulation), which is currently the most frequently used and most effective clinical method for the reduction of the symptoms of PD. In Sect. 6 a comparison of Part I and Part II of The Subthalamic Nucleus is given.

2
The Basal Ganglia

The basal ganglia (BG) are known to play a role in motor, cognitive and associative functions (see Sect. 1 of Part I of *The Subthalamic Nucleus*). However, the role of the BG in motor control appears to be the dominant one, and will be the emphasis of this research. It is widely accepted that the BG play a crucial role in the control of voluntary movement. However, what exactly the BG do for voluntary movement is still under debate. Many clues as to the function of this complex group of subcortical structures have been obtained by examining the deficits that occur following disorders of the BG such as PD and Huntington's disease. Animal models of MPTP-induced parkinsonism have played a crucial role in investigations. Single-cell microelectrode recordings of neuronal activity, as well as imaging studies of blood flow and metabolism have also been employed in an effort to understand the complex interactions between BG nuclei during the execution of movement in normal and parkinsonian subjects. However, despite extensive research on the subject, the function of the BG within the cortico-BG-thalamocortical circuit is still unclear.

The roles attributed to the BG fall into two general categories: those that are directly related to the production of movement, namely focussed selection and inhibition of competing programmes, movement gating and velocity regulation, and action selection; and those that are based on the strengthening of corticostriatal synapses, which may play a role in the learning of movement sequences, namely sequence generation and reinforcement learning.

2.1
Pathways Within the Basal Ganglia

The major pathways within the basal ganglia–thalamocortical circuit, which are known to be involved in the execution of voluntary movement, are illustrated in Fig. 1. The classic view of the pathways through the BG was first proposed by Alexander et al. 1986. According to these authors, two major connections link the BG input nucleus (striatum) to the output nuclei (globus pallidus internus [GPi] and substantia nigra pars reticulate [SNr]), namely the direct and indirect pathways. Normal motor behaviour depends on a critical balance between these two pathways. The BG output nuclei have a high rate of spontaneous discharge and thus exert a tonic, GABA-mediated, inhibitory effect on their target nuclei in the thalamus. The inhibitory outflow is differentially modulated by the direct and indirect pathways, which have opposing effects on the BG output nuclei, and thus on the thalamic targets of these nuclei.

2.1.1
Direct Pathway

The direct pathway arises from inhibitory striatal efferents that contain both GABA and substance P and projects directly to the output nuclei. It is transiently activated by increased phasic excitatory input from the substantia nigra pars compacta (SNc) to the striatum. Activation of the direct pathway briefly suppresses the tonically active inhibitory neurons of the output nuclei, disinhibiting the thalamus and thus increasing thalamocortical activity.

2.1.2
Indirect Pathway

The indirect pathway arises from inhibitory striatal efferents that contain both GABA and enkephalin. These striatal neurons project to the globus pallidus externus (GPe). The GPe projects to the STN, via a purely GABAergic pathway, which finally

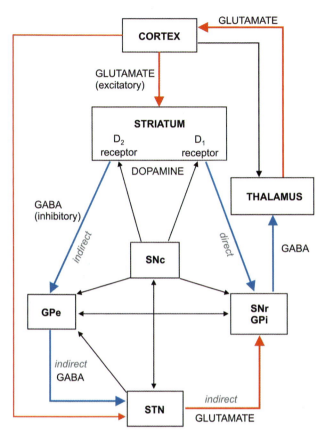

Fig. 1 Cortico-basal ganglia thalamocortical circuit illustrating the direct, indirect and hyperdirect pathways. The STN is located in the indirect and hyperdirect pathway. Excitatory glutamatergic and inhibitory GABAergic projections are indicated. The projection from SNc to the striatum uses dopamine (DA) as neurotransmitter and is inhibitory or excitatory depending on the striatal neurons to which it projects

projects to the output nuclei via an excitatory, glutamatergic projection. There is also a direct projection from the GPe to the output nuclei. The indirect pathway is phasically activated by decreased inhibitory input from the SNc to the striatum, causing an increase in striatal output along the indirect pathway. Normally the high spontaneous discharge rate of GPe neurons exerts a tonic inhibitory influence on the STN. Activation of the indirect pathway tends to suppress the activity of GPe neurons, disinhibiting the STN, and increasing the excitatory drive on the output nuclei. The decreased GPe activity also directly disinhibits the output nuclei. The resulting increase in activity of the output nuclei inhibits the thalamus further, decreasing thalamocortical activity (Fig. 1).

2.1.3
Hyperdirect Pathway

The cortico–STN–GPi hyperdirect pathway has recently received a great deal of attention (Nambu et al. 2000, 2002, 2005; Brown et al. 2003; BarGad et al. 2003; Squire et al. 2003). The hyperdirect pathway conveys powerful excitatory effects from the motor-related cortical areas to the globus pallidus, bypassing the striatum. The hyperdirect pathway is therefore an alternative direct cortical link to the BG, possibly as important to motor control as the corticostriatal–GPi pathway, which is typically considered to be the main cortical relay in the BG. However, the data of the present authors indicate that the cortico-subthalamic connection is far less prominent than the pallido-subthalamic connection.

Anatomical studies have shown that STN-pallidal fibres arborize more widely and terminate on more proximal neuronal elements of the pallidum than striato-pallidal fibres. However, in Part I, Sect. 5.2.4, a point-to-point relation of STN, GPe and GPi groups is favoured. Thus, the striatal and STN inputs to GPi form a pattern of fast, widespread, divergent excitation from the STN, and a slower, focussed, convergent inhibition from the striatum (Squire et al. 2003). Furthermore, cortico-STN neurons and cortico-striatal neurons belong to distinct populations. Thus, signals through the hyperdirect pathway may broadly inhibit motor programmes; then signals through the direct pathway may adjust the selected motor programme according to the situation.

2.1.4
Role of the Direct, Indirect, and Hyperdirect Pathways

During the execution of specific motor acts, movement-related neurons within the BG output nuclei may show either phasic increases or phasic decreases in their normally high rates of spontaneous discharge. Voluntary movements are normally associated with a graded phasic reduction of GPi discharge mediated by the direct pathway, disinhibiting the thalamus and thereby gating or facilitating cortically initiated movements. Phasic increases in GPi discharge may have the opposite effect (Alexander et al. 1990).

There is still debate as to the exact role of the direct and indirect pathways in the control of movement. Two hypotheses have been put forward (Alexander et al. 1990):

1. The scaling hypothesis: both the direct and indirect inputs to the BG output nuclei may be directed to the same set of GPi neurons, whereby the temporal interplay between the activity of direct and indirect inputs allows the BG to influence the characteristics of movements as they are carried out. With this arrangement, the direct pathway facilitates movement, and then, after a delay, the indirect pathway brakes or smoothes the same cortically initiated motor pattern that was being reinforced by the direct pathway.

2. The focusing hypothesis: the direct and indirect inputs associated with a particular motor pattern could be directed to separate sets of GPi neurons. In this configuration, the motor circuit would play a role in reinforcing the currently selected pattern via the direct pathway and suppressing potentially conflicting patterns via the indirect pathway. Overall, this could result in the focussing of neural activity underlying each cortically initiated movement in a centre-surround fashion, favouring intended and preventing unwanted movements.

Nambu et al. (2000, 2002, 2005) propose a dynamic centre-surround model based on the hyperdirect, direct and indirect pathways to explain the role of the BG in the execution of voluntary movement. When a voluntary movement is about to be initiated by cortical mechanisms, a corollary signal is transmitted from the motor cortex to the GPi through the hyperdirect pathway, activates GPi neurons and thereby suppresses large areas of the thalamus and cerebral cortex that are related to both the selected motor programme and other competing programmes (Fig. 2A, top). Next, another corollary signal through the direct pathway is conveyed to the GPi, inhibiting a specific population of pallidal neurons in the centre area, resulting in the disinhibition of their targets and release of the selected motor programme (Fig. 2A, middle). Finally, a third corollary signal through the indirect pathway reaches the GPi, activating neurons therein and suppressing their targets in the thalamus and cerebral cortex extensively (Fig. 2A, bottom). This sequential information processing ensures only the selected motor programme is initiated, executed and terminated at the selected timing.

During voluntary limb movements, the GPi displays an increase in activity in the majority of neurons, with movement-related increases tending to occur earlier

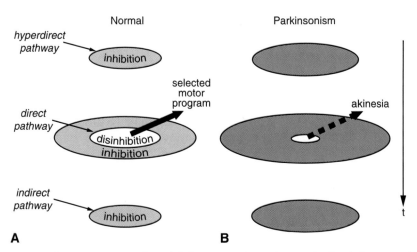

Fig. 2 Dynamic centre-surround model showing activity changes in the thalamus and/or cortex following sequential inputs from the hyperdirect (*top*), direct (*middle*) and indirect (*bottom*) pathways in both the normal case A and the parkinsonian case B. (Source: Nambu et al. 2005)

than decreases. In addition, onset of activity in the STN occurs earlier than that in the pallidum (Mink 1996). Based on these observations, it is likely that the increased pallidal activity during voluntary limb movements is mediated by the net excitatory, faster hyperdirect pathway, while the decreased pallidal activity is mediated by the net inhibitory, slower direct pathway. Pallidal neurons with increased activity may represent those in the surrounding area of the selected motor programme, while pallidal neurons with decreased activity may represent those in the centre area, whose number should be much smaller than that in the surrounding area.

2.1.5
Role of Dopamine in the Direct and Indirect Pathways

Nigrostriatal dopamine projections exert contrasting effects on the direct and indirect pathways. Dopamine is released from the SNc into the synaptic cleft, where it binds to the receptors of the striatum. The effect of dopamine is determined by the type of receptor to which it binds. Striatal neurons projecting in the direct pathway have D1 dopamine receptors which cause excitatory post-synaptic potentials, thereby producing a net excitatory effect on striatal neurons of the direct pathway. Those projecting in the indirect pathway have D2 receptors which cause inhibitory postsynaptic potentials, thereby producing a net inhibitory effect on striatal neurons of the indirect pathway. The facilitation of transmission along the direct pathway and suppression of transmission along the indirect pathway leads to the same effect – reducing inhibition of the thalamocortical neurons and thus facilitating movements initiated in the cortex. Thus, the overall influence of dopamine within the striatum may be to reinforce the activation of the particular basal ganglia-thalamocortical circuit which has been initiated by the cortex.

2.1.6
Conduction Times of Pathways

The time required for a cortical signal to propagate through the BG to the GPi depends on the route taken. It has been observed that the excitatory hyperdirect pathway through the STN (5–8 ms) is faster than the inhibitory direct route through the striatum (15–20 ms) (Squire et al. 2003). Suri et al. (1997) cite a conduction delay through the direct pathway of 20 ms and through the indirect pathway of 30 ms. Furthermore, Romo et al. (1992) describe conduction times of less than 20 ms from striatum to GPi, less than 2.5 ms from GPi to thalamus and less than 4 ms from thalamus to cortex.

2.2
Parkinson's Disease

PD is a molecular disease of the nervous system caused by a defect in dopamine (DA) transmitter metabolism. Eighty percent of the brain's DA is contained in the basal ganglia. The primary pathological feature of PD is a progressive degeneration of midbrain dopaminergic neurons in the SNc. Neuron cell loss of the SNc is remarkably selective, being most severe in the ventrolateral SNc, the remainder of the nucleus being relatively spared (Hassler 1938; Fearnley and Lees 1991; Gibb and Lees 1994; Braak et al. 1996; for a broad review see Usunoff et al. 2002). The affected area of the SNc gives rise to most of the dopaminergic innervation of the sensorimotor region of the putamen. Thus DA loss mainly affects the nigrostriatal pathway. However, other basal ganglia structures, such as the globus pallidus and the STN, also require DA for their physiological activities (Calabresi et al. 2000; see also Sect. 2.3.4.1 of Part I of *The Subthalamic Nucleus*).

The severe dopaminergic degeneration of the SNc results in denervation of dopaminergic terminals in the striatum. The resulting loss of DA-mediated control of striatal neuronal activity leads to an abnormal activity of striatal neurons, which is generally considered to be the origin of PD motor symptoms. DA modulates the activity of striatal cells, and thereby has the potential to modulate the entire circuitry of the basal ganglia. Therefore, to understand how the degeneration of dopaminergic neurons of the SNc can account for the myriad of motor symptoms exhibited by PD, it is necessary to examine the cascade of functional changes which are triggered within all components of the basal ganglia circuitry as a result of DA loss.

2.2.1
Direct and Indirect Pathways in PD

Using the direct/indirect pathway model, PD is explained as an imbalance between the direct and indirect pathways transmitting information from the striatum to the BG output nuclei. The model predicts that the dopaminergic denervation of the striatum leads, ultimately, to an increased firing rate of BG output nuclei, which acts as a brake on the motor cortex, via the inhibitory projection to the thalamus. Due to the differential effects of dopamine on the D1 and D2 dopamine receptors of the striatum, a loss of striatal dopamine results in a reduction in transmission through the direct pathway and an increase in transmission through the indirect pathway. In the direct pathway, a reduction in inhibitory input to the output nuclei occurs. Within the indirect pathway, an excessive inhibition of GPe leads to disinhibition of the STN, which in turn provides excessive excitatory drive to the GPi. The overall effect of such imbalances would lead to increased neuronal discharge in the GPi. The enhanced activity of the output nuclei results in an excessive tonic and phasic inhibition of the motor thalamus. The subsequent reduction of the thalamic glutamatergic output to the motor cortex would cause a reduction in the usual reinforcing influence of the BG motor circuit upon cortically initiated movements. The reduced excitation of the motor cortex might lessen the responsiveness of the motor fields that are engaged by the motor circuit, leading to the hypokinetic symptoms of bradykinesia and akinesia as seen in PD.

2.2.2
Changes in Neuronal Firing Rate in PD

Changes in the neuronal firing rate induced by depletion of striatal DA in PD include increased firing rates in the striatum, GPi and STN and a minimally decreased discharge in the GPe. A summary of studies which back up this conclusion are included below. The tonic firing rates of BG nuclei in the normal and parkinsonian cases are summarized in Table 1.

In PD patients suffering from severe akinesia and rigidity, the spontaneous firing rate of STN neurons was found to be increased to 42.3 ± 22.0 spikes/s, with a range of 10–80 spikes/s (Benazzouz et al. 2002). Similar recordings by Magnin et al. (2000) showed a mean spontaneous firing rate of 41.4 ± 21.3 Hz. These results are supported by other measurements by Magnin et al. (2000) showing mean firing rates of 35 ± 18.8 Hz and 37 ± 17 Hz in parkinsonian patients. STN activity levels in the African green monkey after treatment with MPTP were significantly increased from 18.8 ± 10.3 spikes/s to 25.8 ± 14.9 spikes/s. An even more prominent increase in the firing rate of the 4- to 8-Hz oscillatory neurons was observed. The phasic

Table 1 Tonic firing rates of basal ganglia nuclei

	Tonic activity (Hz) Normal	Parkinsonian	Species	Reference
Striatum	0.1–1		Human	Squire et al. 2003
(projection neurons)		9.8 ± 3.8	Human	Magnin et al. 2000
Striatum	2–10		Human	Squire et al. 2003
(TANs)[1]	5.52		Human	Bennett et al. 1999
STN	20		Human	Squire et al. 2003
		42.3 ± 22.0	Human	Benazzouz et al. 2002
		41.4 ± 21.3	Human	Magnin et al. 2000
		35 ± 18.8		
		37 ± 17		
	18.8 ± 10.3	25.8 ± 14.9	Monkey	Bergman et al. 1994
GPi	60–80		Human	Squire et al. 2003
		91 ± 52.5	Human	Magnin et al. 2000
		89.9 ± 3.0	Human	Tang et al. 2005
	78 ± 26	95 ± 32	Monkey	Filion et al. 1991
	53	60/76	Monkey	Bergman et al. 1994
GPe	70		Human	Squire et al. 2003
		60.8 ± 21.4	Human	Magnin et al. 2000
	76 ± 28	51 ± 27	Monkey	Filion et al. 1991
	62.6 ± 25.8		Monkey	Kita et al. 2004
SNc	2		Human	Squire et al. 2003

[1] TAN, tonically active neurons.

response of STN neurons to the application of flexion and extension torque pulses to the elbow also showed an increase in the average magnitude and duration following MPTP treatment (Bergman et al. 1994).

2.2.3
Changes in Neuronal Firing Pattern in PD

The pattern of discharge of basal ganglia neurons is thought to be equally as important as the rate of discharge in the execution of smooth movements. Several alterations in the discharge pattern have been observed in neurons of the BG in PD subjects, which suggests that the firing pattern may play an important role in the pathophysiology of this disease. These alterations include a tendency of neurons to discharge in bursts, increased correlation and synchronization of discharge between neighbouring neurons, rhythmic and oscillatory behaviour, and a more irregular firing pattern. Increased synchronization of oscillatory discharge appears to be of particular relevance in the development of parkinsonian tremor and may play a role in the development of muscular co-contraction (Wichmann et al. 1996). An abundance of literature exists detailing the changes in firing pattern which occur in the striatum, GPi, STN and thalamus of the PD patient; a summary of the changes in STN is given here.

The percentage of STN cells that discharged in bursts increased from 69% to 79% in African green monkeys following MPTP treatment, and the average burst duration decreased from 121 ± 98 to 81 ± 99 ms (Bergman et al. 1994). Periodic oscillatory activity at low frequency, highly correlated with tremor, was detected in 16% of cells in STN after MPTP treatment, as opposed to 2% before, with an average oscillation frequency of 5.1 Hz (Bergman et al. 1994). Benazzouz et al. (2002), who examined the firing pattern of STN cells of PD patients using single-unit microelectrode recordings, found two types of discharge patterns: a population of cells characterized mainly by tonic activity with an irregular discharge pattern and occasional bursts (mixed pattern) and a population of cells with periodic oscillatory bursts synchronous to resting tremor (burst pattern). Benazzouz et al. (2002) propose that a high level of STN neuronal activity with an irregular and burst-like pattern (mixed pattern) may contribute to akinesia and rigidity, whereas the periodic oscillatory bursts (burst pattern) may contribute to tremor.

2.3
Deep Brain Stimulation

The most effective neurosurgical procedure to date in the treatment of PD is based on the electrical stimulation of small targets in the basal ganglia, a procedure known as deep brain stimulation (DBS). High-frequency electrical stimulation (>100 Hz), with amplitudes ranging from 1 to 5 V and pulse durations of 60–200 µs, is delivered by means of electrodes implanted deep in the brain. A constant stimulation is

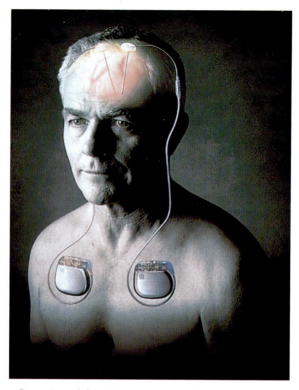

Fig. 3 Overall configuration of the DBS system. The photograph shows the positioning of the implanted bilateral DBS electrodes, the extensions and the pulse generators (pacemakers). (Source: Mogilner et al. 2001)

delivered by a pacemaker-like pulse generator. The positioning of the implanted and pulse generators is illustrated in Fig. 3. The most popular targets for DBS are:

- Motor thalamus (ventral thalamic intermedial nucleus or Vim)
- Globus pallidus internus (GPi)
- Subthalamic nucleus (STN)

DBS of the thalamus is used to treat essential tremor and other forms of tremor. Thalamic DBS can produce an 80% improvement in PD tremor. DBS of the GPi and STN is used to treat the symptoms of PD. Stimulation of these targets has been shown to produce an 80% improvement in PD tremor and dyskinesias, more than 60% improvement in bradykinesia and rigidity, and approximately 40%–50% improvement in gait and postural dysfunction. The GPi or thalamus is targeted for the treatment of dystonia (Lozano et al. 2002).

There is little argument that DBS of the STN, GPi and thalamus has been an effective tool in the treatment of the various symptoms of PD, as well as other movement disorders. However, therapeutic stimulation parameters for DBS (polarity, pulse amplitude, pulse width, frequency) have been derived primarily by trial and error. There remains considerable debate concerning the methods underlying the beneficial effect of DBS and its mechanisms of action are still unknown. Due to the comparable effects of high-frequency stimulation to a lesion of the nucleus, it appears that DBS of the STN or GPi induces a functional inhibition of the stimulated region, and thus to decreased neuronal activity. However, on the basis of physiological principles, one would expect that the effects of DBS are due to excitation of the neural elements (axons, soma) surrounding the tip of the electrode, and thus to increased firing of the axons projecting away from the stimulated region.

Does DBS excite or inhibit its target nucleus? To answer this question it is necessary to examine a number of differential effects which are likely to occur due to stimulation, some or all of which may contribute to the overall observable effect.

2.3.1
Which Neuronal Elements Are Influenced by DBS?

At a physiological level, DBS can have multiple effects on its targets due to the wide range of neuronal elements that may be stimulated by the electrode's current field (Breit et al. 2004; Lozano et al. 2002; Grill et al. 2001). It is known that axons are much more excitable than cell bodies, and that large myelinated fibres are more excitable than unmyelinated axons. Current density decreases with distance from the electrode tip, and axons near the cathode are more likely to be activated than axons near the anode. Electrical stimulation is more likely to activate fibres oriented parallel to the current field than fibres oriented transversely.

Furthermore, electrodes for DBS may be placed in regions with heterogeneous populations of neuronal elements. The applied current may affect several neuronal components in the proximity of the stimulation electrode, with each being subject to both depolarizing and hyperpolarizing effects. Stimulation may influence afferent (axon or axon terminal) and efferent projection neurons, as well as local inter-neurons. Differential effects may occur in the cell body and axon of the same neuron, due to the possibility of a stimulation-induced functional decoupling between cell body and efferent projections (McIntyre et al. 2004a). It was found that the firing of the cell body of directly stimulated neurons is not necessarily representative for the efferent output of the neuron. Extracellular stimulation may also excite or block axons of passage, and fibre activation will result in both antidromic and orthodromic propagation.

2.3.2
Mechanisms of DBS: Hypotheses

Various hypotheses on the mechanisms of action of DBS exist:

- Depolarization block
  High-frequency stimulation may lead to a depolarization block of neuronal transmission by inactivation of voltage-gated sodium and calcium ion channels. A prolonged depolarization of the membrane causes the voltage-gated sodium channels to be trapped in their inactivated state, thus prohibiting the initiation

of new action potentials, inducing a powerful inhibition in the stimulated structure (McIntyre et al. 2004b; Breit et al. 2004; Benabid et al. 2002; Lozano et al. 2002; Grill et al. 2001).
- Activation of afferent inputs
The threshold for activation of axons projecting to the region around the electrode is lower than the threshold for direct activation of local cell bodies. Therefore DBS may excite inhibitory afferent axons projecting to the target nucleus, increasing inhibition of the target and thus playing a role in the suppression of somatic firing. Stimulation may also activate excitatory afferents. The overall effect on the target structure would therefore be the summation of excitatory and inhibitory afferent inputs (McIntyre et al. 2004b; Breit et al. 2004; Benabid et al. 2002; Dostrovsky et al. 2002). In the case of the GPi, DBS may activate inhibitory afferent fibres from the GPe and striatum and excitatory afferent fibres from the STN. As the inhibitory afferents are more numerous, the overall effect is an increased inhibition of the GPi.
- Activation of efferent axons
High-frequency stimulation may activate the efferent projection axons leaving the target structure, directly influencing the output of the stimulated nucleus.
- Synaptic failure
Stimulation-induced synaptic transmission failure may occur due to an inability of the stimulated neurons to follow a rapid train of electrical stimuli. Neurotransmitter depletion or receptor desensitization could result from continuous long-term stimulation. This synaptic depression would lead to the neurons activated by the stimulus train being unable to sustain high-frequency synaptic action on their targets, resulting in reduced efferent output (McIntyre et al. 2004b; Breit et al. 2004; Dostrovsky et al. 2002; Lozano et al. 2002).
- Jamming of abnormal patterns
Stimulation-forced driving of efferent axons (jamming) may impose a high-frequency regular pattern of discharge on the axons, which is time-locked to the stimulation. Insufficient time between DBS pulses may prevent the neurons from returning to their spontaneous baseline activity. DBS disrupts the normal functioning of neurons including any pathological patterns, erasing the burst-like, synchronous firing observed in PD patients, so that the system cannot recognize a pattern. According to this hypothesis, DBS does not reduce neural firing, but instead induces a modulation of pathological network activity, causing network-wide changes (McIntyre et al. 2004b; Breit et al. 2004; Benabid et al. 2002; Garcia et al. 2005a, b; Montgomery et al. 2005).
- Activation of nearby large-fibre systems
Many fibres of passage run close by the structures targeted by DBS. It is possible that direct activation of these fibre tracts may contribute to DBS effectiveness. For example, dopaminergic pathways to the globus pallidus and the striatum pass through the STN, and the axon bundles of pallidothalamic and nigro–thalamic pathways also pass close by (see Part I, Sect. 5.2.1). These pathways may be activated directly by STN stimulation (Grill et al. 2001; Vitek 2002; Miocinovic et al. 2006).

- Neurotransmitter release
  Stimulation may excite axon terminals on the presynaptic neurons which project to the target nucleus. In response to each stimulus, these axon terminals release inhibitory or excitatory neurotransmitters, which diffuse across the synaptic cleft to activate receptors on the target neurons. The release of glutamate induces an excitatory postsynaptic potential (EPSP), whereas the release of GABA induces an inhibitory postsynaptic potential (IPSP) (Grill et al. 2001; Lozano et al. 2002).

  In the case of DBS of the GPi, stimulation may evoke the release of the inhibitory neurotransmitter GABA from the presynaptic terminals of the putamen and GPe, and the excitatory neurotransmitter glutamate from STN neurons. GABAergic synaptic terminals are far more numerous than glutamatergic terminals in the GPi, accounting for approximately 90% of the total synapses (Wu et al. 2001, Boyes and Bolam 2007), so the excitatory effect is masked by the inhibitory effect, resulting in an overall inhibition of the postsynaptic neurons by summation of IPSPs. In contrast, the thalamus contains more excitatory synapses than inhibitory ones, so the effect is one of excitation.
- Antidromic effects
  Electrical stimulation of an axon causes impulses to travel both antidromically as well as orthodromically. Neurons may therefore be activated antidromically via stimulation of their afferent inputs to the target structure. In this way, stimulation of the STN or thalamus could potentially backfire to the cortex by stimulation of cortical inputs to the target structure (Lozano et al. 2002).

In summary, most authors agree that the overall effect of DBS is an inhibition of the target structure, although the stimulation may cause either activation or inhibition of individual neuronal elements in the vicinity of the electrode. Vitek (2002) suggests a possible explanation for the conflicting observations on the effects of DBS; inhibition or excitation. Although DBS may inhibit cellular activity in the stimulated structure via activation of inhibitory afferent fibres projecting to that site, the output from the stimulated structure may be increased, because of the activation of projection axons leaving the target structure, which discharge independently of the soma. This is caused by the possibility of decoupling of the activity of the cell body from the activity of efferent axons. The underlying mechanisms of DBS appear to differ depending on the type of nucleus being stimulated and the exact location of the electrode within the nucleus. The observed effect of stimulation is probably a combination of several of the mechanisms described above. It is important to determine exactly which neuronal elements are affected by DBS in order to obtain a better understanding of the mechanisms by which DBS provides its beneficial effects.

Since only a few studies involve the human STN in neurophysiology (see Benabid 2003) using DBS implanted electrodes and neuroanatomical studies regarding neurodegeneration are only possible post-mortem or by using f-MRI (with a resolution that is too low for neuronal elements), the most reliable approach is modelling the STN.

Investigating the mechanisms of DBS requires knowledge of the stimulation area and the types and number of neuronal elements that are directly stimulated. This is done using volume conduction models (e.g. finite element modelling) in combination with neuronal cell models and network models. In this part, we will concentrate on single-cell modelling and intranuclear and extranuclear modelling.

The final goal is of course to arrive at an optimization of DBS settings and electrode configurations. Questions that arise are:

- Can the nuclear STN circuit be described by a (simple) mathematical model for a contribution to a larger motor basal ganglia–thalamocortical model?
- Can this STN model describe the complex intrinsic and extrinsic activities of the STN under normal and parkinsonian conditions?

Straightforward answers to these questions cannot of course be provided; however, recent literature and research in our Biomedical Signals and Systems group show contributions (see, e.g. Sect. 4, this volume) that partially contribute answers to these questions.

3
STN Activity Recorded in Vitro: Brain Slices

From in vivo experimental studies in monkeys it was found that the neurons of the subthalamic nucleus perform a dual function: (1) they discharge continuously and repetitively at low frequencies (10–30 Hz) in the awake resting state and (2) they discharge bursts of high-frequency spikes (up to several hundred per second), which can last up to 100 ms before, during, and after limb or eye movements in the awake state. The latter activity patterns in turn increase the activity of basal ganglia output neurons and are likely to suppress non-selected motor programmes or terminate sequences of motor behaviour.

In normal monkeys, lesioning of the STN, pharmacologically blocking STN activity or high-frequency stimulation produces a hyperkinetic syndrome (Beurrier et al. 1999). However, in MPTP-treated monkeys, which model Parkinson's disease, these methods show a reduction in motor impairment. In order to gain more insight into the regulation of the basal ganglia by the subthalamic nucleus, it is essential to investigate the underlying ionic mechanisms. In vitro brain slice studies as well as dissociated cell cultures provide a means to (partly) open the closed-loop system of the basal ganglia and to determine the types of membrane channels that are responsible for the regulating behaviour of the STN.

Nakanishi et al. (1987) describe one of the first intracellular recordings from STN neurons in brain slices. Later, a number of investigators added new information by focussing on the plateau potential and low-frequency spikes that were recorded in brain slices and/or acutely dissociated STN neurons. A summary of the results of several studies is described in this section. Details of culturing procedures can be found in the references given.

3.1
Spontaneous Activity

3.1.1
Single-Spike Mode

Tonic discharges of singles spikes in a regular manner were recorded when no additional inputs were applied (Fig. 4). Spontaneous firing rates of 5–40 spikes/s were recorded. Nakanishi et al. (1987) reported that spontaneous firing occurred at membrane potentials between −40 and −65 mV, while Beurrier et al. (1999) found a mean discharge frequency of 22 Hz at membrane potentials ranging between −35 and −50 mV. The cycle of the resting oscillation of STN neurons consisted of a single action potential lasting 1 ms (Nakanishi et al. 1987), which is followed by an afterhyperpolarization that consists of three phases: (1) a fast afterhyperpolarization, (2) a slow afterhyperpolarization and (3) a subsequent slow-ramp depolarization (Fig. 4B).

Bevan and Wilson (1999) tested whether (glutamatergic) excitatory synaptic transmitter release acting within the slice might be responsible for the spontaneous behaviour by bath application of a combination of APV and DNQX, a N-methyl-D-aspartate (NMDA) receptor antagonist and α-amino-3-hydroxy-5-methyl-4-isoxazole-proprionic acid (AMPA) receptor antagonist, respectively. No decrease in spontaneous firing and periodicity of firing was detected, and therefore recurrent excitatory connections within the STN were ruled out, although such connections have been demonstrated (Kitai et al. 1983).

All researchers reported that spontaneous activity totally disappeared in the presence of TTX, which suppresses Na^+-dependent spikes, indicating that voltage-dependent sodium currents are essential to the oscillatory mechanism. A negative slope conductance over the membrane potential range associated with the resting phase is suggested to underlie the spontaneous behaviour, which indicates that

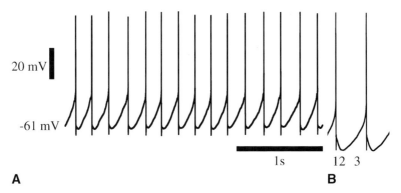

Fig. 4 A STN whole-cell recording showing repetitive spiking. It can clearly be observed that STN cells do not have a stable resting membrane potential. **B** Zooming in on two periods of spontaneous oscillation, the three afterhyperpolarization phases can be distinguished. (Source: Bevan and Wilson 1999)

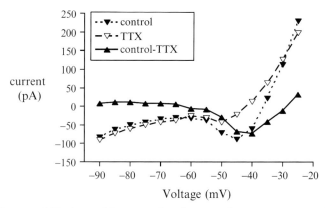

Fig. 5 Steady state I–V plots with a negative slope conductance. In control media, an inward current is activated in the voltage range associated with the depolarization phase of spontaneous activity. Application of TTX suppresses this inward current. Subtracting the TTX curve from the control curve therefore results in the TTX-sensitive sodium current. (Source: Bevan and Wilson 1999)

rhythmic firing is an intrinsic property of STN neurons (Bevan and Wilson 1999). The steady state current-voltage (I–V) curves in Fig. 5 show a negative slope conductance in control media, i.e. an inward current is activated in the voltage range associated with the depolarization phase of spontaneous activity (Bevan and Wilson 1999). This inward current is abolished when TTX is added to the medium.

The large spike afterhyperpolarization suggests the action of a powerful calcium-dependent potassium current. By adding cadmium to the bath, which blocks high-voltage-activated (HVA) calcium currents, spike afterhyperpolarization was drastically reduced in depth and duration, and rhythmic spontaneous firing was disrupted.

The second phase of the afterhyperpolarization suggests the action of a slow hyperpolarization-activated cation current that acts as a time-dependent inward rectifier at very negative membrane potentials (–75 mV or beyond). By adding cesium to the bath, this sag was blocked in a dose-dependent manner (see Fig. 6). This hyperpolarization-activated sag current is, however, not essential for the generation of spontaneous activity since cesium did not affect the spontaneous firing patterns of the cells (Bevan and Wilson 1999).

Apamin treatment, blocking calcium currents, produced a disruption of spike afterhyperpolarization and rhythmic spontaneous firing. As with cadmium, apamin disrupted rhythmic firing only at low rates of activity comparable with those seen spontaneously. Rhythmic firing was still observed at higher (≥ 5 Hz) firing rates. Calcium currents are thus essential for the hyperpolarizing phase of the oscillation, while, as was already suggested by Nakanishi et al. (1987), the depolarization phase of the action potential does not require calcium currents.

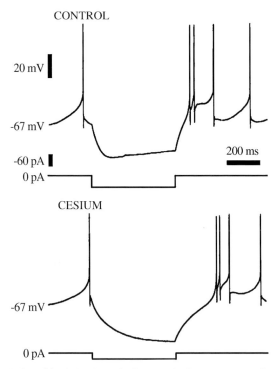

Fig. 6 Sag current induced by injection of a hyperpolarizing current of −60 pA for 500 ms. This sag could be blocked by the presence of cesium; however, cesium did not influence spontaneous activity. (Source: Bevan and Wilson 1999)

A comparison of spontaneous firing at 25°C and 35°C showed that the average frequency of spontaneous firing was approximately twice as fast at the higher temperature (6.5 ± 2 Hz at 35°C, 3.6 ± 1 Hz at 25°C [$n=10$]). Temperature had no effect on the periodicity of firing.

3.1.2
Burst-Firing Mode

Beurrier et al. (1999) found that roughly 46% of the neurons that were examined were also able to fire in bursts while no input was applied (Fig. 7). Burst firing was present in the membrane potential range of −42 to −60 mV, which is somewhat lower than the membrane potentials that were found in neurons showing single-spike activity. Depending on membrane potential, STN neurons were able to switch from one mode to the other.

Two distinct modes of burst firing were detected: (1) pure burst mode, consisting of long-lasting bursts of even duration (ranging from 7 to 29 bursts per minute; bursts lasting longer than 800 ms showed a rapidly increasing spike frequency after

Fig. 7 Burst-firing mode detected using patch-clamp techniques. Burst-firing mode was also recorded using intracellular techniques. (Source: Beurrier et al. 1999)

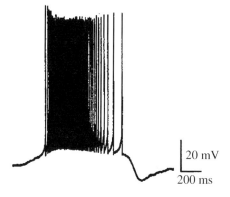

Fig. 8 Burst with increasing and decreasing spike frequency, respectively. The membrane potential slowly repolarizes during the burst. (Source: Beurrier et al. 1999)

the first spike, followed by a gradual decrease, see Fig. 8); (2) mixed burst mode with alternating bursts of long and short duration.

A summary of pharmacological tests on burst activity is given here (Beurrier et al. 1999):

- Drugs interfering with Ca^{2+} entry or intracellular free Ca^{2+} ions all had an inhibitory (irreversible) effect on burst firing.
- Bath application of nifedipine (L-type Ca^{2+} channel blocker), largely reduced the duration of bursts and even suppressed burst firing.
- Nickel (Ni^{2+}) blocking T/R-type Ca^{2+} channels (reversibly) decreased the duration of bursts.
- Bath application of the permeable form of the Ca^{2+} chelator BAPTA-AM largely reduced the duration of bursts and even suppressed burst firing after a delay of about 40 min at all potentials tested.
- Tetraethylammonium chloride (TEA), blocking the large conductance Ca^{2+}-dependent K^+ current, but not the delayed rectifier one, and apamin, which selectively blocks small conductance Ca^{2+}-dependent K^+ current, totally prevented burst repolarization and suppressed spikes.

3.2
Depolarizing and Hyperpolarizing Inputs

At resting membrane potentials, depolarizing current pulses were found to evoke action potentials either during the initial phase of the current or during the entire period of current injection after which the membrane potential returned to rest in an exponential manner (Otsuka et al. 2001). Figure 9 gives an example of recorded action potentials resulting from depolarizing inputs. With increased depolarizing current injection, the firing rate increased up to approximately 500 Hz according to Nakanishi et al. (1988), with an almost linear slope of 900 Hz/nA up to 300 Hz. Bevan and Wilson (1999) found a sigmoidal current–frequency plot (Fig. 10). However, the sigmoidal shape was lost and replaced by an initially linear relationship after treatment with apamin or cadmium, which blocks high-voltage-activated (HVA) Ca^{2+} currents.

The observed linear current–frequency relationship for a certain input range, as found by different experimenters, indicate that STN neurons are extremely sensitive to small changes in their excitatory inputs. According to Bevan and Wilson (1999), the average of the membrane potential trajectory during high-frequency firing moves more deeply into the negative slope region of the steady-state

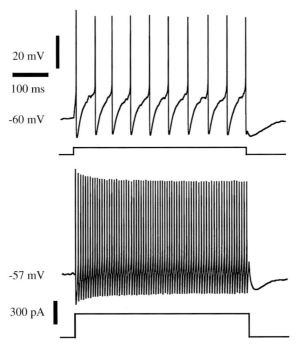

Fig. 9 Depolarizing input currents result in increased firing rates. A small decrease in voltage range is seen at higher firing rates. (Source: Bevan and Wilson 1999)

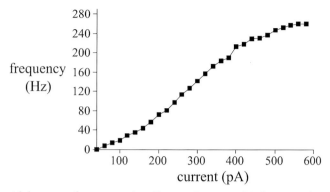

Fig. 10 Sigmoidal current-frequency plot. (Source: Bevan and Wilson 1999)

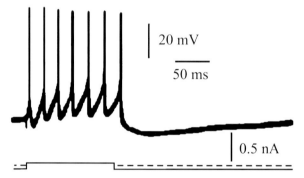

Fig. 11 A depolarizing input pulse induces repetitive spiking with a long-lasting hyperpolarization potential after the termination of the current pulse. Spontaneous firing was eliminated by the continuous injection of a hyperpolarizing current of 0.7 nA. (Source: Nakanishi et al. 1987)

current–voltage curve, which suggests that the persistent sodium current is responsible for the enhanced sensitivity in this linear range. This explanation may also account for the speed-up of firing seen at the onset of high-frequency trains in STN neurons (Fig. 8).

At membrane potentials more depolarized than −30 mV, spike amplitude decreased and spike frequency increased, leading rapidly to a blockade of STN tonic activity (Beurrier et al. 1999). On the other hand, at membrane potentials more hyperpolarized than −60 to −70 mV, cells were also silenced (Beurrier et al. 1999).

Many experiments were conducted in which spontaneous activity was blocked by a hyperpolarizing input. An example is shown in Fig. 11; repetitive firing induced by a depolarizing pulse was followed by long-lasting (250–600 ms) hyperpolarizing potentials with an amplitude of 5–12 mV at the break of a current pulse (Nakanishi et al. 1987). This long-lasting hyperpolarizing potential was diminished by superfusing Ca^{2+}-free medium, indicating that Ca^{2+} channels are involved in this hyperpolarizing phase. In addition, it was not affected by intracellular Cl^- injection.

Nakanishi et al. (1987) observed two distinct TTX-resistant potentials in STN neurons, which were dependent on the membrane potential:

1. Slow depolarizing potential (membrane potentials of −50 to −65 mV), later called a plateau potential (Beurrier et al. 1999; Otsuka et al. 2001; Song et al. 2000). Action potentials could be evoked during this depolarization.
2. Slow action potential (membrane potentials of −65 mV or beyond), later called a low-threshold spike (Beurrier et al. 1999).

These potentials were considered to represent activation of inward Ca^{2+} currents since their generation was blocked by superfusion of Ca^{2+}-free medium or application of Co^{2+}, which is known to block Ca^{2+} conductance.

3.2.1
Plateau Potential

At hyperpolarized states, long-lasting plateau potentials were generated in response to depolarizing or hyperpolarizing current pulses (100 pA, 100 ms) that clearly outlasted the duration of the applied current pulses up to even 500 ms or more (Fig. 12). Otsuka et al. (2001) were the first to use a definition of a plateau potential using half-decay times, the time interval from the pulse end to the time when the potential decayed to half-amplitude at the current pulse end; a potential with a half-decay time 0.2 s or greater is defined as a plateau potential. However, describing a plateau potential as a long-lasting depolarizing potential suggests an elevated membrane potential; nevertheless, the degree to which the membrane should be depolarized is not clearly included in the definition. In the modelling study by Otsuka et al. (2004), no reference is made to the plateau potential definition (see also Sect. 5.1, this volume).

As shown in Fig. 12, the slow depolarization could trigger repetitive firing of action potentials, a burst, in which the firing rate increased along with the development of the slow depolarization. The slow depolarizing potential was TTX-resistant, but was suppressed by superfusion of Ca^{2+}-free medium.

Two phases in the plateau potential can be discerned when TTX is added, suppressing sodium currents (see Fig. 13): (1) a slow depolarization triggered by the

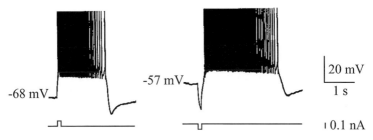

Fig. 12 Plateau potentials recorded in whole-cell configurations, generated by a depolarizing (*left*) and hyperpolarizing (*right*) input current. (Source: Beurrier et al. 1999)

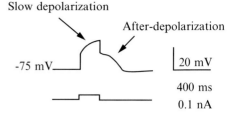

Fig. 13 Plateau generated by a depolarizing current input in the presence of TTX. (Source: Beurrier et al. 1999)

depolarizing current pulse (50 pA, 200 ms) and (2) an afterdepolarization triggered at the break of the current pulse.

Neurons in which a plateau potential could be evoked by a depolarizing current pulse at hyperpolarized states also generated a plateau potential after termination of a hyperpolarizing current pulse. Nevertheless, plateau potentials were triggered within a narrow range of membrane potentials: Beurrier et al. (1999) found a range between −50 and −75 mV; Otsuka et al. (2001) reported a threshold hyperpolarization level at which a plateau potential was first induced of −74.98 ± 1.96 mV. All these results show that plateau potentials are induced by the activation of voltage-dependent conductances. The results from whole-cell recordings using different types of Ca^{2+} channel blockers suggested that both Ca^{2+} entry through L-type Ca^{2+} channels and intracellular free Ca^{2+} ions are involved in the generation of plateau potentials. Beurrier et al. (1999) concluded that the ionic conductances underlying bursts and plateau potentials are the same.

The early phase of the plateau potential was found to be insensitive to membrane perturbations; the stability index, defined as the ratio of the peak potential after a perturbing current pulse during a plateau potential to the potential immediately before the current, was 1 or close to 1 during the initial phase of the plateau potential (Otsuka et al. 2001). This robustness gradually decreased towards the end of the plateau potential, as was tested by the injection of negative current pulses.

In comparison to the influence of temperature on spontaneous firing, the occurrence of action potentials in combination with a plateau potential is also dependent on temperature. A plateau potential that did not evoke action potentials at 20°C did evoke action potentials even at its late phase at 25°C. Raising the temperature also appeared to increase the duration of the plateau potential.

3.2.2
Low-Threshold Spike

Beurrier et al. (1999) observed a small, transient depolarization triggering a few spikes at the break of a short hyperpolarizing current pulse in a subpopulation (71%) of the STN neurons tested. This was called a low-threshold spike (LTS) because of its negative threshold compared with Na^+-dependent spikes. The low-threshold spike

Fig. 14 Three superimposed voltage traces in response to hyperpolarizing current pulses of fixed amplitude (−150 pA) and increasing duration (40, 80 and 120 ms). LTS was evoked in a neuron maintained at −64 mV when the membrane was held at −85 mV for at least 80 ms during the application of a hyperpolarizing current pulse. (Source: Beurrier et al. 1999)

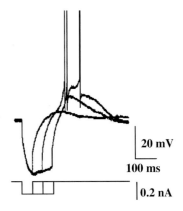

is distinguished from the plateau potential by its fast rate of rise, a short duration (about 30 ms) and a large peak amplitude (Fig. 14). Only in neurons with a membrane potential more negative than −65 mV were these observed, and thus, a Ca^{2+} conductance responsible for the slow action potential is inactivated in the depolarized membrane. A depolarizing sag of the membrane potential indicates the presence of a hyperpolarization-activated cation current (I_h). LTS was unaffected by TTX, but completely disappeared in a Ca^{2+}-free Co^{2+} containing external solution, or in the presence of a low concentration of Ni^{2+}. All these results suggest that a rapid voltage-inactivating, Ni^{2+}-sensitive current such as the low-threshold T/R-type Ca^{2+} current underlies LTS.

According to Beurrier et al. (1999), two populations of STN neurons can be distinguished in vitro: (1) STN neurons that are able to burst and generate LTS and plateau potentials, and (2) STN neurons that only respond with an LTS.

Whether these physiological types of neurons are identical to the two types of STN projection neurons distinguished morphologically (see Sect. 2.1 of Part I) is unclear. According to Otsuka et al. (2001) plateau-generating neurons tend to be located in the lateral part of the nucleus. However, although the morphology of plateau-generating neurons did not appear to differ from that of non-plateau-generating neurons, the input resistance at resting membrane potentials of plateau-generating neurons was found to be significantly larger than that of non-plateau-generating neurons (813 ± 70 vs 524 ± 50 MΩ).

3.3
Ionic Mechanisms of a Plateau Potential

The fact that a plateau potential could be induced only from a hyperpolarized state suggests that voltage-dependent conductances are involved in the generation of the plateau potentials (Otsuka et al. 2001). The occurrence of a plateau potential requires the steady-state current-voltage (I–V) curve to cross zero current with a negative slope. Thus, according to Otsuka et al. 2001, in STN neurons, synaptic potentials have to activate an inward current that decays only slowly for plateau potentials to occur. The channel giving rise to this current has to be inactivated again (or partially activated) at the resting membrane potential and deinactivated at hyperpolarized potentials. The high-threshold L-type Ca^{2+} channel which has slow inactivation dynamics depending on both the membrane potential and Ca^{2+} satisfies this description. Otsuka et al. (2004) (see also Sect. 5, this volume) shows that the voltage dependence of plateau potential induction can be solely attributed to the voltage-dependent inactivation of L-type Ca^{2+} channel. Repolarization of plateau potentials is suggested to be mediated by both Ca^{2+}-dependent K^+ channels and TEA-sensitive K^+ channels.

T-type channels were thought to be important in the generation of oscillatory behaviour; however, according to Song et al. (2000), STN neurons do not appear to generate rhythmic bursting by themselves, in contradiction to the observations of Beurrier et al. (1999) (see Sect. 3.1.2). A possible explanation for the oscillatory bursting activity observed can be found in the involvement of synaptic inputs, since T-type channels in STN neurons have a preferential distribution in dendritic processes (Song et al. 2000; see Sect. 2.3.5 of Part I of *The Subthalamic Nucleus*). This was found from acutely isolated STN neurons in which dendritic processes are lost during dissociation. No low-voltage-activated channels were found in these neurons, suggesting that the low-threshold T-type channels are located within the dendritic processes. Evidence was found by current-clamp recordings in slice preparations; at resting membrane potentials, a hyperpolarizing current always resulted in a slow spike, which was sensitive to micromolar concentrations of Ni^{2+}, suggesting that it is a low-threshold Ca^{2+} spike. In contrast, in acutely dissociated cells activation occurred at around −50 mV, suggesting the activation of high-threshold channels. Furthermore, T-type currents are characterized by fast inactivation (on the order of 25 ms), while the shortest inactivation time constants that were observed in dissociated cells were longer than 200 ms. T-type channels are also suggested to be involved in the generation of low-threshold spikes (LTS), since they have slower kinetics in comparison to Na^+ channels.

Next to the high-voltage-activated L-type Ca^{2+}, and low-voltage-activated T-type Ca^{2+} currents, Song et al. (2000) found that STN neurons in acutely isolated and slice preparations also express the high-voltage-activated N-, Q-, and R-type subtypes.

N-type currents are known to be subject to neuromodulation by a number of receptors coupled to trimeric guanine-binding proteins. N-type currents play a dominant role in STN neurons, which makes these neurons highly modifiable. A variety of neurotransmitters are known to be released within the STN, including GABA, acetylcholine, serotonin, glutamate, and probably dopamine (see Sect. 2.2 of Part I). These may all modulate Ca^{2+} entry through N-type channels, depending on the receptors expressed. In addition, this modulation may also change the activity of Ca^{2+}-dependent processes such as Ca^{2+}-dependent K^+ channels.

For the different subtypes of high-threshold Ca^{2+} currents, Song et al. (2000) determined Boltzmann fitted curves with the following median half-activation voltages:

- N-type, −18.6 mV (identified by its sensitivity to ω-CgTx).
- L-type, −19.8 mV (identified by its sensitivity to dihydropyridines).
- Q-type, −15.9 mV (identified by its sensitivity to high concentrations of ω-AgTx).
- R-type, −18.6 mV (this type has the lowest threshold; its presence is based on the fact that saturating blockers to other subtypes of current were used, since R-type current is only defined by its resistance to all known organic blockers).

STN neurons express a Kv3-type delayed rectifier and an A-type current as depolarization-activated K$^+$ channels. The high threshold of activation of Kv3 channels and the fast inactivation of I_A probably causes a high-input resistance at potentials close to resting membrane potentials, enabling the generation of a plateau potential by small amounts of inward currents. This would confirm the high input resistance at resting membrane potentials as observed for plateau-generating STN neurons in contrast with non-plateau potential-generating neurons (see Sect. 3.2, this volume), suggesting a lack or a different distribution of A-type channels in the latter type of STN neurons.

3.4
Synaptic Inputs

The activity of a neuron is determined by its intrinsic membrane properties, but also by extrinsic synaptic inputs. STN receives excitatory inputs from cortex and thalamus, and inhibitory inputs from globus pallidus. From the previous section, it may be suggested that STN cells work as linear transformers relaying excitatory inputs to their targets by increasing their firing rate with increasing depolarizing input currents. STN neurons, however, have intrinsic membrane properties that can significantly change the neuronal firing pattern, i.e. the generation of a plateau potential. Because of this plateau potential, which has slow decay kinetics, high-frequency bursting may occur even in the absence of synaptic inputs.

The interaction of intrinsic membrane properties of STN neurons with their synaptic inputs was investigated by Otsuka et al. (2001). In this study, brain slices including STN, obtained from Sprague-Dawley rats (14–20 postnatal days) were stimulated at a location rostral to the STN (axon collaterals of fibres descending in the cerebral peduncle enter the STN from the rostral aspect; see Part I, Sect. 5.2.1) in order to evoke synaptic potentials in the STN, i.e. EPSPs possibly of cortical origin. Bicuculline was included in the external solution to block inhibitory synaptic potentials. In response to stimulation, depolarizing potentials or inward currents were observed in most of the neurons examined. It was determined that these potentials are EPSPs of monosynaptic nature mediated by glutamate (AMPA and NMDA; see also Part I, Sect. 2.3.4.4) receptors. With increased stimulus strength, the EPSPs always triggered a single-action potential in STN neurons at resting membrane potentials. However, when the membrane potential was hyperpolarized to approximately −75 mV, stimulation with the same strength evoked a plateau potential in a subpopulation of STN neurons, in a voltage-dependent manner. On the rising phase of the plateau potential, a short train of action potentials (1–5 spikes) was evoked.

Excitatory synaptic inputs to STN neurons are electrotonically distant from the soma. Voltage-clamping the soma, as performed in brain slices, still allows the plateau potential to be evoked in dendrites by the synaptic input, which would result in an inward current similar in shape to the plateau potential to be recorded in the soma. Since these currents were never observed, plateau potentials in STN neurons appear to be generated in the soma, which is consistent with previous findings that L-type Ca^{2+} channels tend to be localized in the soma (Otsuka et al. 2001). Furthermore, those neurons that could generate a plateau potential by current injection could also generate plateau potentials by EPSPs, suggesting common membrane mechanisms playing a role in the two situations.

As concluded in the previous section, for rhythmic bursting activity, as described by Beurrier et al. (1999), to occur in STN cells, it is expected, as investigated in explant cultures, that (part of) the STN-GP network is intact (Plenz and Kitai 1999). In this situation, two features of STN plateau potentials may be relevant:

1. Because a plateau potential can be evoked as a rebound potential, a short train of spikes in GP neurons would hyperpolarize STN neurons and a plateau potential would then occur as a rebound potential, evoking a train of spikes in STN neurons.
2. STN activity would cause immediate feedback inhibition from the GP, but this inhibition might not immediately terminate STN spiking activity because the early part of plateau potentials appears to be resistant to inhibitory perturbations.

In this case, a hyperpolarization of the STN membrane, as required for a plateau potential to occur, is caused by inhibitory inputs from GP neurons. Another option for the generation of a hyperpolarization may be the opening of K^+ channels by metabolic signalling pathways (Otsuka et al. 2001).

3.5
High-Frequency Stimulation of STN Cells

To gain more insight into the mechanisms of DBS, it is of interest to investigate the response of STN cells to high-frequency stimulation. As shown in Sect. 3.2, the ability of STN neurons to fire at very high frequencies in response to depolarizing input current suggests that they are unlikely to show a substantial depolarization block. A summary of the results of several studies follows.

Beurrier et al. (2000) observed that stimulating subthalamic neurons in brain slices for 1 min using 100-µs bipolar stimuli at a frequency ranging between 100 and 250 Hz completely blocked spontaneous activity. The effects are found to be independent of synaptic activity and are mediated by blocking the persistent Na^+ current as well as the L- and T-type Ca^{2+} currents.

Benazzouz et al. (1995, 2000a, b) found that after high-frequency stimulation (130 Hz) in STN in rats, activity of cells in pallidal areas as well as near the stimulation site in STN was suppressed in comparison to the activity recorded before stimulation. Artefacts made it difficult to establish how activity changes during stimulation. It is suggested that STN activity may be suppressed by the activation of a calcium-gated potassium afterhyperpolarization current.

Garcia et al. (2003) determined the effect of low- (<30 Hz) and high-frequency stimulation of the STN on the electrophysiological activity of STN neurons in naive and dopamine-depleted slices. A summary of the results, which were similar for both types of slices (pulse durations of 60 or 90 µs were used) is provided immediately below:

- At 10 Hz spontaneous activity was not suppressed for amplitudes up to about 1.5 mA;
- At 30 Hz, spontaneous activity was suppressed in a subset of cells.
- At 50 Hz, single spikes at 50 Hz, trains of spikes at 50 Hz separated by short periods of silence, or bursts of 50 Hz spikes.
- At 80–185 Hz and intensities of 400–800 µA, spontaneous activity was always suppressed; however, cells were not silenced, but entirely driven by the stimulation. Recurrent bursts, either mixed or with single spikes were evoked.

Furthermore, it was observed that after the stimulation was stopped, all neurons became immediately and transiently silent for 20 s up to 8 min, while membrane potentials were approximately −69.7 mV in naive slices and −66.5 mV in dopamine-depleted slices.

The effect of high-frequency stimulation is assumed to have a postsynaptic origin, because it was still present when glutamatergic, GABAergic, and aminergic synaptic transmission was suppressed (Garcia et al. 2003). As also proposed by Beurrier et al. (2001), Na$^+$ channels in the cell body or axon initial segments are thus directly activated by stimulation, resulting in a depolarization of the membrane, which results in activation of L-type Ca^{2+} channels, allowing Ca^{2+} to enter the cell and inducing a bursting activity. In addition, Garcia et al. (2003) expect that stimulation of presynaptic aminergic (dopaminergic) or serotoninergic afferents to the STN do not play a major role in the mechanisms of high-frequency stimulation in STN.

Montgomery and Baker (2000) proposed an augmentation of neurotransmitter release resulting from the high-frequency stimulation driving the efferent axons from STN. In normal rats, they observed that high-frequency stimulation in STN increased extracellular glutamate in GPi and SNr, which receives excitatory input from STN. Increases in blood oxygenation in subcortical regions during high-frequency stimulation of the STN suggest a high activity in STN targets.

3.6
Intrinsic Versus Extrinsic Properties: Bursts

3.6.1
Definition of Bursts

Wilson et al. (2004) have shown that whilst it is possible pharmacologically to induce low-frequency, oscillatory burst firing in the STN in vitro, this is not accompanied by synchrony. Even more, it is suggested that the circuitry needed to produce synchrony in the STN is not intrinsic to the STN itself but requires connections with other basal ganglia nuclei. Therefore, bursting activity is an inherent property of (STN) neurons, while STN synchrony is dependent on the extrinsic connections. Burst in vitro may therefore be a reaction to disturbing the existing connections in vivo, for example, caused by dissociation, lesions, or the artificial environment; they are artificial. On the other hand, bursts in vivo with intact connections are a meaningful reaction to induce certain functions (e.g. sleep). They can be meaningful in learning or be used for target reaching.

Synchronous bursting may also be defined as network bursting in which a number of action potentials are generated within a certain time frame by different cells. All in all, what is a burst and what is its definition?

"A burst is a train of closely spaced action potentials. It is convenient to treat doublets (two spikes) and triplets (three spikes) as short bursts. Bursting can be generated by a strong synaptic input, which drives the neuron over the threshold and makes it fire repetitive action potentials. Alternatively bursting can be generated by an intrinsic voltage or Ca^{2+}-gated ionic mechanism, which may be triggered by a brief synaptic input. In this case, the bursting pattern is stereotypical, usually with a constant or slowly decreasing interspike (intraburst) frequency" (Izhikevich et al. 2003). This description holds for the in vivo situation.

In principal two action potentials within a small time distance make a burst.

- What is the determined small time period? The scientist is free to choose. One can find all kind of definitions in the literature.
- Therefore, the definition of a burst is dependent on the scientist's estimation of the time between successive action potentials.
- Moreover, few scientists accept that a doublet or triplet is a burst; but how many then are needed: four, five, or more?

Furthermore, reducing the concentration of Mg^{2+} can induce bursting (Shen 1996). For studies that use dissociated neurons in culture or those working with slices of brain tissue, the ion concentration in the fluids used is immanent. Therefore it is hard to compare experimental studies, and in order to use experimental data to determine model parameters, the average of several experiments is required.

3.6.2
Burst Detection Algorithms

For the study of recorded action potentials using multi-electrode arrays (MEA), various definitions are used. A difference is made between bursts on one electrode and those on several electrodes, called synchronized or network bursts. Network bursts can have a time scale from 100 ms to 4 s. For its study time stamp signals are transformed into time discrete signals by studying the number of spikes per time interval. This method reduces the resolution, which can be improved after detection. Moreover, the duration of the interval has an enormous influence on the results (Segev et al. 2004; Eytan and Marom 2006). A series of approaches will be treated hereafter:

Van Elburg and van Ooyen (2004) consider a burst being two or more short intervals, followed by a long interval. They developed a measure for bursting, called B_2:

$$B_2 = \frac{2Var(t_{i+1} - t_i) - Var(t_{i+2} - t_i)}{2(E(t_{i+1} - t_i))^2}$$

It is based on a data set, and cannot be used to calculate the time periods of individual bursts.

Van Pelt et al. (2004) are interested in network bursts (synchronized bursts). To localize these bursts, a calculation of a (temporary) increase in the number of spikes and the number of active electrodes is made. Per 100-ms bin, they analysed the product of the number of electrodes which contain a spike and the total number of spikes, called the burst intensity product (BIP). If this product rises above a threshold, the presence of a network burst is accepted. Network bursts contain consecutive bins. A start and an end of a burst can be found. This calculation depends on the number of active electrodes in culture. Per channel clusters of spikes are traced by looking for regions where spike intervals are shorter than 1.3 s.

Wagenaar et al. (2005a, b) investigated a measure of the number of bursts present in a recording. A 5-min recording is subdivided into 300 periods or 1-s bins. Successively, the number of spikes present in 45 (15%) of the most active bins can be detected. The burst index (BI) is calculated as follows:

$$f_{15} = \frac{\#\ spikes\ in\ 15\%\ most\ active bins}{total\ \#\ spikes}, BI = \frac{f_{15} - 0.15}{0.85}$$

The BI is in between 0 and 1. These 45 bins (15%) were chosen because the most active (bursting) cultures contained 45 bins in which bursts could be detected.

MEABench-software (Wagenaar et al. 2005) detects so-called burstlets on each electrode. A burstlet consists of 4 consecutive spikes with an inter-spike interval (ISI) smaller than a threshold value $\tau_{c,core}$:

$$\tau_{c,core} = \begin{cases} \frac{1}{4f} &, \frac{1}{4f} \leq 100\,ms \\ 100\,ms, & otherwise \end{cases}$$

$$\tau_{c,lobes} = \begin{cases} \frac{1}{3f} &, \frac{1}{3f} \leq 200\,ms \\ 200\,ms, & otherwise \end{cases}$$

After detection with threshold $\tau_{c,core}$, burstlets are extended with ISIs with a lower threshold, $\tau_{c,lobes}$. A burst consists of a series of burstlets that overlap in time. Post-processing detects two burstlets that are integrated into one.

Tam (2002) gives an algorithm that detects bursts per channel based on ISIs. In the formula, $B_{i,k}$ is the duration of a burst and I_i is the ith ISI.

$$B_{i,K} = \sum_{k=i+1}^{j} I_k \text{ if } \sum_{k=i+1}^{j} I_k < I_i \text{ and } \sum_{k=i+1}^{j} I_k < I_{j+1}$$

A burst is detected if the sum of j ISIs is smaller than the ISI before and after the burst.

3.6.3
Network Bursts Using Burst and Phase Profiles

Stegenga et al. (2007) refer to bursts as network bursts, when the total firing rate, as determined in 10-ms bins, crosses a threshold. The threshold is set at 2 spikes for each electrode that showed spiking activity during a measurement. An electrode was considered active if its spike rate exceeded 0.1 Hz.

Whenever a bin exceeded the threshold, a burst profile was calculated in order to estimate the time at which the peak of an array-wide spiking rate (AWSR) occurred. Once found, burst and phase profiles were calculated from t − 300 ms to t + 300 ms. To ensure that there was no overlap between profiles, all bins in t − 600 to t + 600 ms were set to 0. Thereafter, two profiles are made: a burst profile and a phase profile.

A burst profile is an estimation of the instantaneous AWSR. To this end, all the spike occurrences in a burst are taken together and convolved with a Gaussian probability density function (Fig. 15). The standard deviation is 5 ms, which was wide enough to give a smooth result near the maximum AWSR, and small enough so as not to obscure all details of the AWSR. The burst profiles can be compared to each other by aligning them to the maximum instantaneous firing rate. The profiles are 600 ms wide, which was large enough to capture the relevant features.

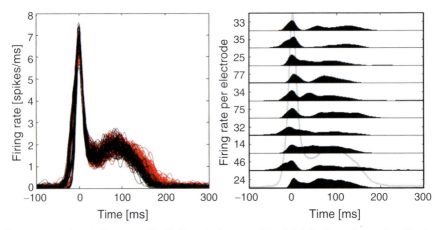

Fig. 15 Example of a burst profile (*left*) and phase profiles (*right*) of ten electrodes. Shining through the phase profiles is the total sum: the burst profile

The number of spikes that are generated per electrode per burst is relatively small, when compared to the number of spikes in the AWSR. Therefore, more averaging in the time domain is required. This can be done by increasing the width of the Gaussian inter polation function, or by including several aligned bursts. It was chosen to include multiple bursts because a wider Gaussian would obscure the very details one is looking for.

4
STN Activity Recorded in Vitro: Dissociated Cell Cultures

In PD, because of the loss of the dopaminergic nigrostriatal connection, the basal ganglia neurons, including the STN, fire by low-frequency oscillatory bursts and fire in synchrony. Under normal conditions, the basal ganglia neurons fire irregularly. The reciprocal pallido–subthalamic connection can be mimicked in vitro and is capable of burst firing. This pattern can be influenced by co-cultures of striatum and cortex (Plenz and Kitai 1999). The observed bursting activity was unaccompanied by synchronous activity in STN slice preparations after the addition of glutamate, dopamine, GABA, or muscarine receptor agonists and/or antagonists. Therefore, synchrony of STN activity in the parkinsonian state is dependent on its extrinsic connections (Wilson et al. 2004).

Dissociation of central nervous system areas of P1 (day 1 postnatal) rat pups makes it possible to culture these neurons in a chemically defined medium (Heida 2003). By placing these cultures on MEAs, their spontaneous electrical activity can be recorded. By using polyethylenimine as substrate, a seemingly monolayered network can be created in culture on the MEA (see Rutten et al. 2001 for an overview).

4.1
Experimental Set-up

In short, microelectrode arrays could be constructed because of the progress in the field of microtechnology and were applied in the biological sciences for cardiac cells (Thomas et al. 1972); later it became a tool in neurotechnology for recording and stimulation purposes (Gross et al. 1979). Until now, the MEA that has been used for stimulation and recording experiments with neuronal cells and tissue at the Twente University consists of a glass substrate (5×5 cm) on top of which an array of electrodes is created using lithographic and etching methods (Multi Channel Systems). Figure 16 shows the layout of the MEA. Around the centre, a glass ring provides a culture chamber for neurons suspended in a specially prepared culture medium (Romijn et al. 1984).

If dissociated subthalamic neurons are placed on a MEA they are devoid of their connections and after approximately 7 days, when a network has been created, their activity can be recorded.

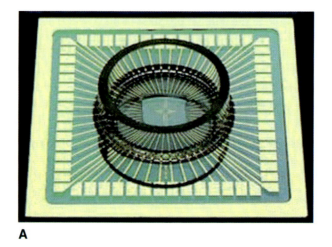

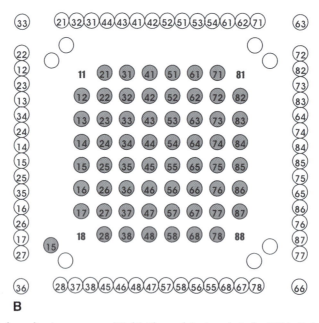

Fig. 16 MEA for culturing neurons (Multi Channel Systems); **A** the MEA; **B** the inner electrode structure in detail and electrode numbering. Electrode diameter is 10 μm; inter-electrode distance is 100 μm

From Part I, Sects. 2.2 and 5.2, it is clear that a cholinergic innervation of the STN area is present. From the presence of N-type Ca^{2+} channels, as described in Sect. 3.3 (this volume), we may expect that acetylcholine can change the activity of STN neurons. Therefore, adding acetylcholine is a mode to test the activity of the STN neurons in terms of increased or decreased firing rate, changes in the firing pattern, or inducing or changing burst activity behaviour in culture.

4.1.1
Cell Culture

Subthalamic areas were dissected from rat pups (postnatal day 1, P_1), mechanically dissociated and trypsin-treated and cultured in a serum-free medium (R12, Romijn et al. 1984), with NGF added, for at least 10 days in vitro (DIV) at a concentration of nearly 10^4 subthalamic cells/ml. An extended description of the technique can be found in van Welsum et al. (1989) and van Dorp et al. (1990).

4.1.2
Measurement Set-up

A MC1060BC preamplifier and FA60s filter amplifier (both Multi Channel Systems) was used to prepare the signals for AD conversion. Amplification is 1,000 times in a range from 100 Hz to 6,000 Hz. A 6024E data-acquisition card (National Instruments, Austin, TX, USA) was used to record all 60 channels at 16 kHz (Figure 17 shows the measurement set-up). Custom-made Labview (National Instruments) programmes are used to control the data acquisition (DAQ). These programmes also apply a threshold detection scheme with the objective of data reduction. Actual detection of action potentials is performed in an offline fashion. During the experiments, the temperature was controlled at 36.0°C using a TC01 (Multi Channel System) temperature controller. Recording starts after a minimum of 20 min to prevent any transient effects. Noise levels were typically 3–5 μV_{RMS}, somewhat depending on the MEA and electrode. We use commercially available MEAs from Multi Channel Systems with 60 titanium nitride electrodes in a square grid. The inter-electrode distance is 100 µm, and the diameter of the electrodes is 10 µm.

4.2
Spontaneous Activity

Spontaneous activity was observed using several MEAs for a total of several hours. Figure 18 shows the result of one of the measurements in terms of the average spike rate as a function of time over a period of 5 min. Electrodes that showed a minimum firing activity of at least 1 Hz were selected for further analysis, which in this case resulted in the selection of 22 electrodes. The average firing rate in Fig. 18

Fig. 17 Measurement set-up in which the MEA is connected to the computer while culturing conditions are controlled

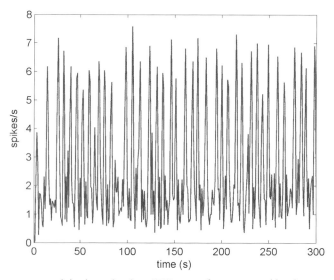

Fig. 18 Spontaneous activity in an in vitro STN network represented by the average number of spikes as a function of time. An average spiking frequency of 2.7 Hz was detected selecting those electrodes with at least 1 Hz as a baseline activity

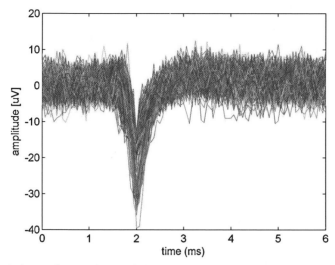

Fig. 19 Typical waveforms of recorded STN spontaneous spiking activity; this graph consists of 500 randomly selected wave forms from one of the electrodes

is 2.7 Hz. However, the average firing was found to vary among MEAs, possibly resulting from different network architectures. A number of wave forms from the measurement are shown in Fig. 19.

4.3
Addition of Acetylcholine

To test the effect of acetylcholine on the activity of STN cells, acetylcholine was added to a concentration of 10 µM in 50 µl of culture medium to a 2-ml culture bath, each 1,000 ms. The total recording lasted 1.5 h; the recorded activity before addition of acetylcholine was considered the spontaneous STN activity in culture. Whether bursts occur in the network of STN cells was tested. A burst was defined as consisting of at least four action potentials with an inter-spike interval of 20 ms or less, i.e. a minimum intra-burst spike rate of 50 Hz.

Addition of acetylcholine induced a direct substantial reduction of spontaneous activity in the cultures, as shown in Figs. 20 and 21. At the application, for a time period of nearly 50–100 s no activity was noticed, after which network activity returned. Measuring the total spike activity over the entire period, a 25% reduction of activity occurred as compared to the spontaneous control activity before acetylcholine addition (3.1 and 4.2 Hz, respectively). In a control experiment, it was shown that the addition of 50 µl R12 (normal culture medium) without acetylcholine did not disturb the activity at all.

Bursting activity was not convincingly detected in the networks. Only one of the MEAs showed bursting activity in accordance with the burst definition, with

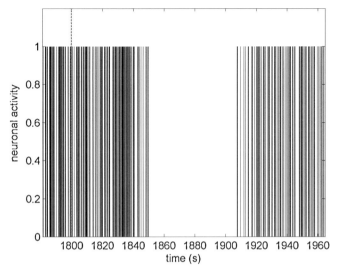

Fig. 20 Activity recorded at electrode 14 during and after the addition of acetylcholine (step 1, addition of 10 µMol ACh). The data clearly show a total reduction of activity after the moment ACh was added

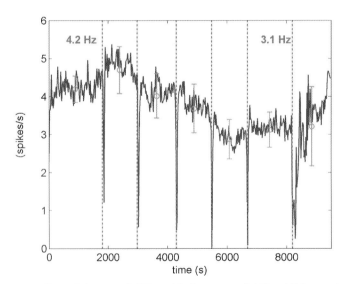

Fig. 21 Average spike activity over 8,000 s with five steps of ACh addition and a washing step after 8,100 ss, indicated by the broken lines. After the removal of ACh, network activity returned to the normal level of spontaneous activity

an average burst length of 4.2 spikes. This bursting activity did not change significantly before and during a period of 1,000 s after the addition of acetylcholine.

From these results, we can conclude that the addition of 50 µM acetylcholine to cultured STN area neurons shows two effects:

1. A direct strong reducing effect on the spike activity of the STN neurons after each addition step that lasts about 50–100 seconds.
2. A long-term effect reducing the activity with, in this case 25% as compared to the spontaneous activity in culture.

The connection that is mimicked by addition of acetylcholine is part of the PPN–STN connection (see Part I, Sect. 5.2.5). This part of the PPN–STN connection is cholinergic, but other cell groups are also involved (glutamatergic, and possibly GABA-ergic). Destruction of the PPN ends up with hyperactivity of the STN (Breit et al. 2005). PPN lesioning was shown to induce akinesia (in fact only motor hypoactivity) in primates (Matsumura and Kojima 2001; Matsumura 2001). It is well established that the cholinergic agonists brought into the rat STN contribute to an excitation of the STN neurons (Feger et al. 1979). However, it was also shown that muscarine agonists in slices diminished the amplitude of both EPSPs and IPSPs in the STN (Flores et al. 1996; Shen and Johnson 2000). The reduction of IPSPs is higher, which leads to a final excitation of STN neurons (Rosales et al. 1994; Shen and Johnson 2000).

Contradictory results are found in the literature as to the effect of acetylcholine on the subthalamic neurons (see above). This could well be caused by the still existing connections. Taking away one connection by lesioning, adding neurotransmitters or their agonists, therefore, does not show the pure effect of connections, neurotransmitters or receptors. Too many parameters are involved to fully understand the effect of these experiments. Culturing subthalamic neurons at least restricts the amount of parameters, but adds others!

It is rather unexpected that addition of acetylcholine to such cultures shows a short-term and a long-term effect. If hyperactivity of STN is induced by reducing the PPN neurotransmitters, among them acetylcholine, and motor hypoactivity is the consequence, then this MEA culturing experiment explains by the long-term effect how such hyperactivity can result from this type of neurotransmitter, neglecting all the other effects of other PPN neurotransmitters. One should remember that addition experiments of Plenz and Kitai (1999) do not show any effects on synchrony. Our results did not show any effect on bursting activity, which may suggest that the long-term effect of acetylcholine on subthalamic cultured cells is related to this synchrony or pacemaker effect, stressing the role of the PPN.

4.4
Electrical Stimulation

Electrical stimulation through one of the electrodes of the multi-electrode array was applied at 20 Hz and 80 Hz with the following stimulation and measurement settings:

- 20 Hz, 500 charge-balanced block pulses (+10/−10 µA, 400 µs/phase), start at 300 s (end 325 s).
- 80 Hz, 2,000 charge-balanced block pulses (+10/−10 µA, 400 µs/phase), start at 300 s (end 325 s).

Stimulation artefacts were removed from the recorded data. Electrodes with average spontaneous activity of at least 1 Hz prior to stimulation were selected.

The recorded data show that at low-frequency (20 Hz) as well as at high-frequency stimulation (80 Hz) the average firing rate increased right after the onset of stimulation, Figures 22 and 23, respectively. During the stimulation period, the firing rate rapidly decreased in both situations; however, at high frequency (80 Hz) this decrease was more rapid. At the time that stimulation is switched off, network activity is quickly diminished before regaining its spontaneous activity. This silencing effect coincides with the effects observed by Benazzouz et al. (2000) and Garcia et al. (2003), as described in Sect. 3.5 (this volume).

All observed effects were found to be reproducible. In Fig. 23B, stimulation at 80 Hz is applied for 120 s, which shows that during stimulation network activity stabilizes and the deviation from the average spike rate is decreased, suggesting a more regulated network activity.

The dissociated STN cells form a network in vitro, while in vivo the STN does not show recurrent connections. No definite conclusions on the stimulation experiments in relation to the use of high-frequency stimulation in DBS can be

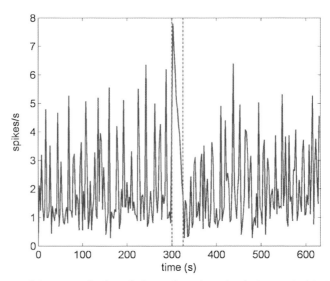

Fig. 22 Average firing rate of selected electrodes. The stimulation period is indicated by broken lines; stimulation frequency is 20 Hz, and a total of 500 pulses are applied (i.e. 25 s). Electrode 28 is the stimulation site

Electrical Stimulation

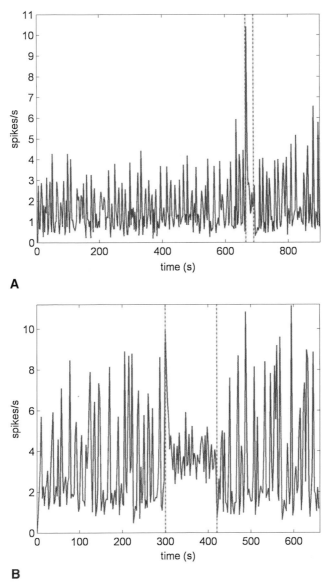

Fig. 23 A,B The effect of electrical stimulation at 80 Hz on the average firing rate of STN network activity; the stimulation period is indicated by broken lines. **A** A total of 2000 pulses is applied (i.e. 25 s); **B** an extended period of stimulation (120 s) indicates that network activity is regulated by stimulation. In both situations, electrode 28 is the stimulation site

drawn. Further research including the use of different types of neurotransmitters and stimulation protocols is required. Modelling studies as described in Sect. 5 may also shed some light on the network dynamics and the influence of stimulation.

5
STN Cell Models and Simulation of Neuronal Networks

Neurons can be modelled using the properties of the ion channels in the membrane. There are several methods that use Hodgkin and Huxley type equations (Hodgkin and Huxley 1952) to describe ion channel dynamics. Heldoorn et al. (2001a, b) produced Onuf's nucleus neurons to explore the possibilities of sphincter contractions. In this paper, the parameters for developing the activity of an Onuf neuron had to be partially borrowed from the motor neuron of the spinal cord. In fact, the required output to manage an artificial anal sphincter was such that an explanation for incontinence could be brought forward. However, some of the parameters had to be estimated, since no evident data were available.

The same holds for the STN: limited information regarding the presence, types and properties of ion channels in human STN neurons is available. Therefore STN models are mostly based on information obtained from studies on rats. A comparison between two single-compartment models of STN neurons is described in view of their spontaneous activity and the transformation of this activity into a bursting pattern as observed under parkinsonian conditions (dopamine depletion) and recorded in brain slices (see Sect. 3).

As described in Sect. 3.2.1, a plateau potential can be generated in STN neurons when these cells are hyperpolarized in advance. By means of the voltage-dependent generation of a plateau potential, STN cells can induce long-lasting high-frequency discharge in the absence of synaptic inputs. This ability is thought to play important roles in the generation of oscillatory bursting activity of the STN neurons, as observed in PD. It is assumed that the hyperpolarized state results from dopamine depletion, and thus bursting activities are more likely to be induced under parkinsonian conditions than in the normal situation, since normally dopamine depolarizes STN neurons such that it prevents GP inhibitory inputs to hyperpolarize them enough for the generation of plateau potentials and thus bursting activity.

5.1
Otsuka's Model

The model of Otsuka et al. (2004) is based on the dynamics involved in the voltage-dependent generation of a plateau potential. According to Otsuka et al. (2004), single-compartment models are justified because experimental studies suggested that the subcellular origin of a plateau potential (the cause of bursting activity) is the soma and/or proximal dendrites. The responses of the model to injection of depolarizing current pulses at the resting and hyperpolarized membrane potentials were compared with recordings from plateau-generating neurons in brain slices.

5.1.1
Membrane Dynamics

The membrane potential v of the single compartment model of the STN is described by:

$$C_m \frac{dv}{dt} = -I_{Na} - I_K - I_A - I_T - I_L - I_{Ca-K} - I_{leak}$$

in which:

$I_{Na} = g_{Na} m^3 h (v - v_{Na})$ Na⁺ current, with activation variable m and inactivation variable h;

$I_K = g_K n^4 (v - v_K)$ delayed rectifier K⁺ current (high activation threshold, fast activation time constant), with activation variable n;

$I_A = g_A a^2 b (v - v_K)$ A-type K⁺ current (low activation threshold, fast activation and inactivation time constants), with activation variable a and inactivation variable b;

$I_T = g_T p^2 q (v - v_{Ca})$ low-threshold T-type Ca²⁺ current with activation variable p and inactivation variable q;

$I_L = g_L c^2 d_1 d_2 (v - v_{Ca})$ L-type Ca²⁺ current with activation variable c, voltage-dependent inactivation variable d_1, and Ca²⁺-dependent inactivation variable d_2.

$v_{Ca} = \frac{RT}{zF} \frac{[Ca^{2+}]_{ex}}{[Ca^{2+}]_{in}}$ Nernst equation for calcium, with $[Ca^{2+}]_i$ the intracellular calcium concentration, $[Ca^{2+}]_{ex}$ the extracellular calcium concentration (2 mM), R the gas constant, T absolute temperature (of which no indication is given by Otsuka et al. (2004) other than a temperature of 30°C during the experiments), and z the valence, which in this case is 2; reversal potentials of other ionic channels were assumed constant;

$I_{Ca-K} = g_{Ca-K} r^2 (v - sv_K)$ Ca²⁺-activated K⁺ current with Ca²⁺-dependent inactivation variable r;

$I_{leak} = g_{leak} (v - v_{leak})$ leak current.

$\frac{d[Ca]_i}{dt} = -\alpha I_{Ca} - K_{Ca} [Ca]_i$ intracellular Ca²⁺ concentration, depends on the total Ca²⁺ current, K_{Ca} is the removal rate (ms⁻¹), $\alpha = \frac{1}{zF}$ with z the valence of calcium, and F Faraday's constant. Note that without any correction for the membrane area, the dimension of the first term on the right side of this equation seems to be incorrect.

C_m is the membrane capacitance and is set at 1 µF/cm². Currents are expressed in µA/cm²; conductances are expressed in mS/cm².

As indicated by Song et al. (2000), N-, P-, and Q-type calcium channels are not included since these channels have not been found to be involved in plateau potentials.

The types and dynamics of the ionic channels were identified by patch clamp and whole cell recordings from slices (Otsuka et al. 2000; Beurrier et al. 1999; Song et al. 2000; Wigmore and Lacey 2000; Do and Bean 2003; Rudy and McBain 2001). Gating kinetics of the ionic conductances were calculated using the following equations:

$$\frac{dw}{dt} = \frac{w_\infty - w}{\tau_w}, \text{ with } w = a, b, c, d_1, d_2, h, m, n, p, q, r$$

Steady state activation and inactivation functions are:

$$w_\infty = \frac{1}{1 + \exp[(v - \theta_w)/k_w]}$$

with θ_w and k_w the half-inactivation/activation voltage and slope, respectively.

The activation time constants used are:

$$\tau_w = \tau_w^o + \frac{\tau_w^1}{1 + \exp[(v - \theta_w^\tau)/\sigma_w]}$$

for $w = a, m$

$$\tau_w = \tau_w^o + \frac{\tau_w^1}{\exp[-(v - \theta_w^{\tau 1})/\sigma_w^1] + \exp[-(v - \theta_w^{\tau 2})/\sigma_w^2]}$$

for $w = b, c, d_1, d_2, h, n, p, q, r$

Parameter values can be found in appendix 1. Figures 24 and 25 show the steady state activation and inactivation variables I_L, I_T and I_{Ca-K}. Figure 26 shows the steady state time constants of these variables; the time constants for the Ca^{2+}-dependent variables are constant (τ_{d_2}=130 ms; τ_r= 2 ms).

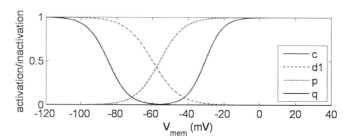

Fig. 24 Steady-state voltage-dependent activation and inactivation variables: activation c and inactivation variable d_1 of the L-type Ca^{2+} current I_L; activation p, and inactivation variable q of the low-threshold T-type Ca^{2+} current

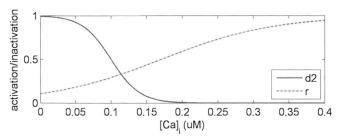

Fig. 25 Steady-state Ca^{2+}-dependent activation and inactivation variables: activation r of the Ca^{2+}-activated K$^+$ current, inactivation d_2 of the L-type Ca^{2+} current, I_L

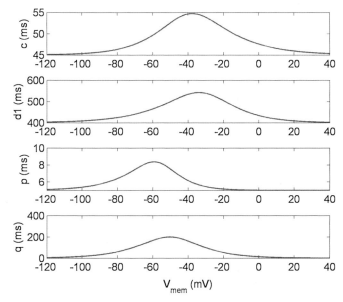

Fig. 26 Steady-state time constants of the voltage-dependent activation and inactivation variables: τc of activation variable c, τ_{d_1} of inactivation variable d_1 of the L-type Ca$_{2+}$ current I_L; τ_p of activation variable p, and τ_q of inactivation variable q of the low-threshold T-type Ca^{2+} current

5.1.2
Spontaneous Activity

The reproduction of the model of the STN neuron without additional inputs showed a spontaneous spiking rate of about 5 Hz (Fig. 27). According to Otsuka et al. (2004), the model neuron fires at about 10 Hz, while from Fig. 1A in the paper, a frequency of about 6 Hz can be estimated. The produced wave form of a single action potential (Fig. 28) is similar to the one presented by Otsuka et al. (2004).

The duration of a single action potential is approximately 2 ms; the resting membrane potential is roughly −58 mV with membrane potentials, varying from −65 to 40 mV during action potentials. Comparing the spontaneous activity with Fig. 4

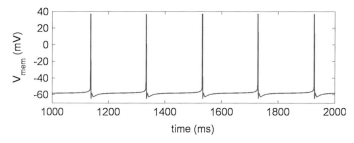

Fig. 27 Spontaneous activity of the modelled STN neuron. (Reproduced from Otsuka et al. 2004)

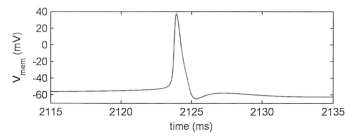

Fig. 28 Membrane potential during a single action potential (spontaneous activity). (Reproduced from Otsuka et al. 2004)

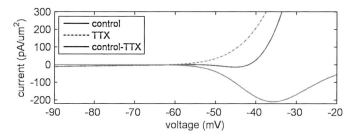

Fig. 29 Negative slope conductance resulting from a TTX-sensitive sodium current

(Sect. 3.1, this volume), the fast afterhyperpolarization can be discerned followed by a slow afterhyperpolarization phase; however, the membrane subsequently remains at around a resting membrane potential of −57 mV, in contrast to the slow-ramp depolarization (the third phase), as observed by Bevan and Wilson (1999).

Figure 29 shows the steady-state I–V curves for membrane potentials ranging from −90 to −20 mV, for the normal (control) situation, and while Na$^+$ currents are blocked (TTX, corresponding to blocking sodium channels by the application of TTX, as performed in experimental studies). Subtracting the TTX curve from the control curve results in a (TTX-sensitive) sodium current with a negative slope

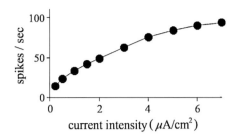

Fig. 30 Current–frequency relationship of the STN model. (Source: Otsuka et al. 2004)

conductance, as was suggested to underlie spontaneous activity, as described by Bevan and Wilson (1999; see Sect. 3.1).

With depolarizing input currents, the firing rate of the STN neuron model increases. The relationship between the firing rate with the magnitude of depolarizing input currents is shown in Fig. 30 and is qualitatively comparable to the results from in vitro brain slice experiments; however, the maximum frequency was approximately 100 Hz, while experimental data show frequencies up to 500 Hz (Sect. 3.2, this volume).

A constant hyperpolarizing input current (≤ -1.5 μA/cm^2) silences spontaneous activity of the STN cell (not shown).

5.1.3
Plateau Potential Generation

The model is able to produce plateau potentials and burst firing when from a hyperpolarized state, a depolarizing input current is applied (Fig. 31). For membrane potentials below approximately −70 mV, a plateau potential was induced with burst spiking that outlasted the current injection in comparison to experimental data (lower two graphs in Fig. 31). The duration of the plateau potential, defined as the duration from the pulse end to the time when the membrane potential returned to the baseline level, is comparable to experiments. However, according to the definition of Otsuka et al. (2001), plateau potentials should have a minimum half-decay time of 200 ms. According to this definition, none of the responses presented can be termed plateau potentials. However, Otsuka et al. (2004) do not use this definition. A clear long-term elevation of membrane potential during the bursting activity can be discerned, which will, in the following sections, be used to indicate the presence of a plateau potential.

A hyperpolarizing current pulse is able to induce burst firing in the neuron model. Figure 32 gives an example of a burst resulting from a hyperpolarizing pulse of −5 μA/cm^2 that is applied for 300 ms. No additional inputs were applied, and after the burst the neuron regains its spontaneous activity. A gradual decrease in the firing rate is observed during the last phase of the burst in comparison to the experimental observations of Beurrier et al. (1999). No clear long-lasting depolarizing potential is seen in this situation, which would indicate that this is not a

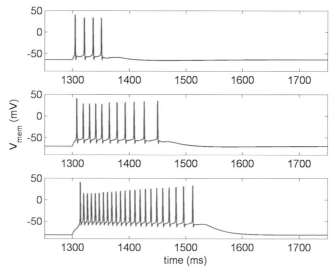

Fig. 31 Plateau potential generation with action potentials resulting from a depolarizing input current of 5 μA/cm², for 50 ms, from $t=1{,}300$ to $t=1{,}350$ ms, while the membrane is kept at a hyperpolarized state obtained by continuous application of a hyperpolarizing current of −2, −4, and −7.5 μA/cm², respectively. In the *upper graph*, the membrane has not been sufficiently hyperpolarized in order to induce a plateau potential

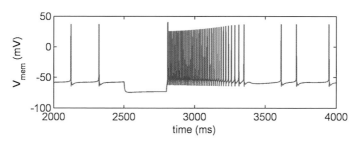

Fig. 32 A hyperpolarizing pulse of −5 μA/cm² is applied for 300 ms; starting at $t=2{,}500$ ms at resting membrane potential induces burst firing

plateau potential. This result also contradicts with the experimental result as presented in Fig. 4 of Otsuka et al. (2004) and the description of the model result (Fig. 4B of Otsuka et al. (2004)), and Fig. 12 of Sect. 3.2.1 (Beurrier et al. 1999).

At the break of a hyperpolarizing input current, while in addition a constant hyperpolarizing current is applied which maintains the membrane at a hyperpolarized state, a clear elevation of the membrane potential, i.e. a plateau potential, is induced at the break of the pulse in combination with the generation of a burst, as shown in Fig. 33.

The mechanisms underlying plateau potentials can be investigated by blocking Na⁺ channels ($g_{Na}=0$), which abolishes spiking activity. Figure 34A and B shows the

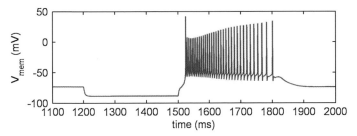

Fig. 33 Plateau potential generation with burst activity resulting from an input current of −5 μA/cm2 for 300 ms (from $t=1{,}200$ to $t=1{,}500$ ms), while the membrane is kept at a hyperpolarized state obtained by continuous application of a hyperpolarizing current of −5 μA/cm2

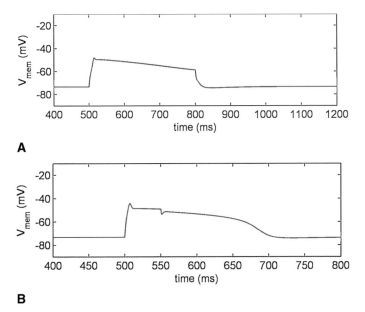

Fig. 34 **A** The response to a depolarizing input pulse of 5 μA/cm^2 with a pulse width of 300 ms, starting at $t=500$ ms, does not outlast the duration of the pulse. **B** The response to a depolarizing input pulse of 10 μA/cm^2 and duration of 50 ms, starting at $t=500$ ms results in a plateau potential. In both situations, sodium channels are blocked and the membrane is kept at a hyperpolarized state by a constant input of −5 μA/cm^2

response of the STN neuron model to a depolarizing current pulse of 300 and 50 ms, respectively. In both situations, the membrane is hyperpolarized to approximately −72 mV. Although a clear depolarization is observed in Fig. 34A, the response does not outlast the current pulse. In contrast, Fig. 34B shows a depolarization that lasts beyond the duration of the applied pulse.

In addition, Fig. 35A and B shows the response of an STN cell to hyperpolarizing pulses; in A no hyperpolarized state is attained and no long-lasting bursting activity is generated in contrast to B.

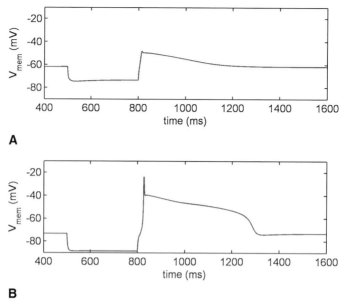

Fig. 35 **A** The response to a hyperpolarizing input pulse of −5 µA/cm² and 300 ms starting at t = 500 ms and **B** the response to that same hyperpolarizing input pulse while the membrane is kept at a hyperpolarized state by a constant input of −5 µA/cm². In this case, a plateau potential is generated. In both situations, sodium channels are blocked

Comparing the activation and inactivation mechanisms of the different ion channels may reveal the currents responsible for the induced responses. Activation and inactivation variables corresponding to the simulations of Fig. 35A are presented in Fig. 36. Likewise, activation and inactivation variables corresponding to the simulation of Fig. 35B are presented in Fig. 37.

From a comparison of Figs. 36 and 37, we may conclude that deinactivation of the T-type current ($q \rightarrow 0.6$) during the current pulse, and the activation of the T-type current ($p \sim 1$) and deinactivation of I_{CaK} ($r \sim 1$) at the break of the current pulse with slow deactivation and inactivation, respectively, seem to be responsible for the generation of a plateau potential. Furthermore, the small activation of L-type current (c) after the break of the current pulse, and the more prominent role of the inactivation of I_{CaK} (b) during and after the current pulse as well as the slow dynamics of inactivation of the L-type current (d_1 and d_2), may add to the generation and extent of the plateau potential.

In order to test the role of the L-type Ca^{2+} current, Otsuka et al. (2004) applied a slow voltage ramp to the model, while the Na^+ conductance was set to zero. Only when the model was held at hyperpolarized potentials before the ramp did the currents that were evoked show a negative slope. Figure 38 shows the result; in this case, negative currents were obtained when the membrane was initially brought to

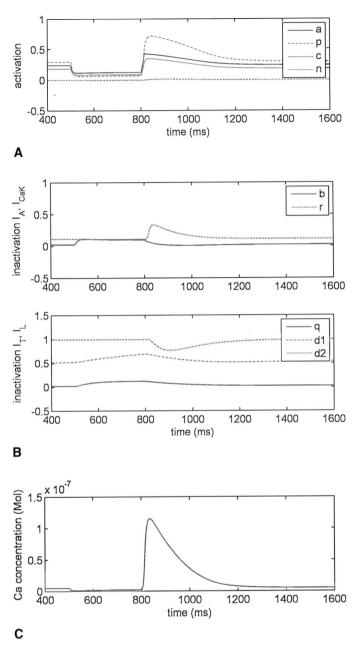

Fig. 36 A–C Details of the response to a hyperpolarizing input pulse of $-5\ \mu A/cm^2$ and 300 ms starting at $t = 500$ ms while sodium channels are blocked (Fig. 35A). **A** Activation variables a, c, and p of current I_A, I_T, I_L, respectively. **B** Inactivation variables b and r of current I_A and I_{CaK}, respectively (*upper graph*), q of the current I_T, and d_1 and d_2 of current I_L (*lower graph*). **C** Internal Ca^{2+} concentration

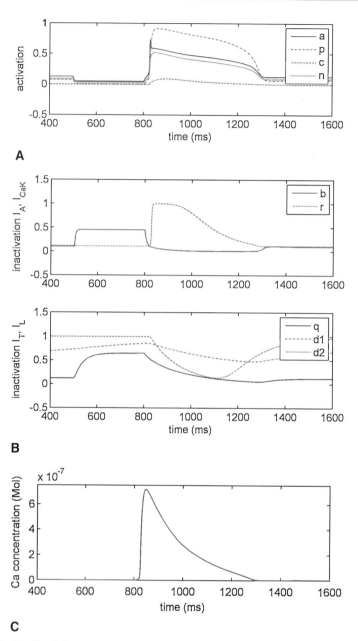

Fig. 37 A–C Details of the response to a hyperpolarizing input pulse of −5 µA/cm² and 300 ms starting at t = 500 ms while sodium channels are blocked and the membrane is continuously hyperpolarized by a constant input of −5 µA/cm² (Fig. 35B). **A** Activation variables a, c, and p of current I_A, I_T, I_L, respectively. **B** Inactivation variables b and r of current I_A and I_{CaK}, respectively (*upper graph*), q of the current I_T, and d_1 and d_2 of current I_L (*lower graph*). **C** Internal Ca²⁺ concentration

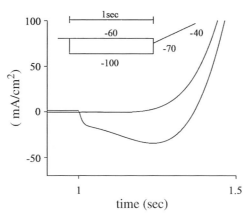

Fig. 38 Simulated current–voltage relations with sodium conductance, g_{Na}, set to 0 (re-presenting addition of TTX). Ramp-shape voltage change preceded by a 1-s prepulse, which sets the membrane potential at −60 or −100 mV (schematically shown in the *inset* in the figure). A negative slope region is obtained when the pre-pulse sets the membrane at −100 mV. (Source: Otsuka et al. 2004)

−100 mV using a pre-pulse. It was found that the inactivation variable of the L-type Ca^{2+} current allows the model to change the shape of the steady-state I–V curve in a voltage-dependent manner. Removal of the inactivation variable c resulted in a voltage independent generation of a plateau potential, as reported by Otsuka et al. (2004); however, no results were presented.

Otsuka et al. (2004) tested the model according to two pharmacological experiments:

- TEA: decreasing the value of maximal delayed-rectifier type K^+ conductance increases the duration of plateau potentials.
- Ca^{2+} chelation by BAPTA according to: $\dfrac{d[Ca]_i}{dt} = -\dfrac{\alpha I_{Ca}}{X} - K_{Ca}[Ca]$ with X the rate of chelation; the duration of the plateau potential was sensitive to the rate of Ca^{2+} chelation. Increasing X increases the duration of the plateau potential.

Although the STN model is able to show firing behaviours resembling those observed in physiological experiments, it does not constitute a physiologically complete description of STN neurons. The focus was on the generation of plateau potentials, and therefore ionic currents having less obvious relationships with plateau potentials were not included. Conductances that have been physiologically identified in STN neurons, but were not included in the model are:

- TTX-sensitive sustained Na^+ current (I_{NaP}), which can impart a negative slope region to the steady-state I–V curve of STN neurons (Beurrier et al. 2000; Bevan and Wilson 1999).

- Hyperpolarization-activated inward current (I_h) responsible for the sag, and which may be involved in the repolarization kinetics of a plateau potential in addition to the inactivation of L-type Ca^{2+} channels and the activation of Ca^{2+}-dependent K^+ channels (Beurrier et al. 1999; Song et al. 2000).
- Several types of high-threshold currents; according to Song et al. 2000, L-type channels have the lowest activation voltage in comparison to the other high-threshold Ca^{2+} channels in STN. Otsuka et al. (2004) used −25 mV for the half-inactivation voltage. In acutely dissociated preparations, Ba^{2+} was present, and since activation and inactivation properties of Ca^{2+} depend on species and concentration of charge carriers, exact physiological values of Ca^{2+} channels remain unknown. It is therefore not known which subtypes of L-type channels are expressed in STN neurons.

5.2
Terman and Rubin's Model

The STN model of Terman et al. (2002) is used for the investigation of the mechanism of oscillatory activity in the subthalamo-pallidal network, which was suggested by Plenz and Kitai (1999). In Sect. 5.6 (this volume), more details of this network activity are described. In the following sections, the dynamics of the STN neuron model that is part of this network are highlighted.

5.2.1
Membrane Dynamics

The membrane potential v of the single-compartment model of the STN according to Terman and Rubin is described by:

$$C_m \frac{dv}{dt} = -I_{Na} - I_K - I_{Ca} - I_T - I_{AHP} - I_{leak}$$

in which the incorporated ionic currents are described as follows:

$I_{Na} = g_{Na} m_\infty^3 h(v - v_{Na})$ Na^+ current, with instantaneous activation variable m_∞ and inactivation variable h;

$I_K = g_K n^4 (v - v_K)$ delayed rectifier K^+ current (high activation threshold, fast activation time constant), with activation variable n;

$I_{Ca} = g_{Ca} s_\infty^2 (v - v_{Ca})$ high-threshold Ca^{2+} current with instantaneous activation variable s_∞;

$I_T = g_T a_\infty^3 b_\infty^2 (v - v_{Ca})$ low-threshold T-type Ca^{2+} current, with instantaneous activation variable a_∞ and inactivation variable b_∞; by using this equation the T-type current includes the effects of a hyperpolarization-activated inward current, the sag;

$I_{AHP} = g_{AHP}(v - v_K) \dfrac{[Ca^{2+}]_{in}}{[Ca^{2+}]_{in} + k_1}$ Ca^{2+}-activated, voltage-independent afterhyperpolarization K^+ current, with $[Ca^{2+}]_{in}$ the intracellular concentration of Ca^{2+} ions, and k_1 the dissociation constant of this current;

$l_{leak} = g_{leak}(v - v_{leak})$ leak current.

The gating kinetics of the ionic conductances were calculated using the following equations:

$$\frac{dw}{dt} = \frac{w_\infty - w}{\tau_w}$$

with $w = n, h, r$

The steady state activation and inactivation functions are:

$$w_\infty = \frac{1}{1 + \exp\left[-(v - \theta_w)/k_w\right]}$$

with $w = n, m, h, a, r, s$, and θ_w and k_w the half-inactivation/activation voltage and slope, respectively.

The inactivation function b_∞ of the T-type current is determined as follows:

$$b_\infty = \frac{1}{1 + \exp[(r - \theta_b)/k_b]} - \frac{1}{1 + \exp[-\theta_b/k_b]}$$

The activation time constants used are:

$$\tau_w = \tau_w^0 + \frac{\tau_w^1}{1 + \exp\left[-(v - \theta_w^\tau)/\sigma_w\right]}$$

for $w = n, h, r$

The intracellular Ca^{2+} concentration is determined by:

$$\frac{d}{dt}\left[Ca^{2+}\right]_{in} = \varepsilon\left(-I_{Ca} - I_T - k_{Ca}\left[Ca^{2+}\right]_{in}\right)$$

in which the constant ε combines the effects of buffers, cell volume, and the molar charge of calcium; k_{Ca} is the calcium pump rate constant. All currents are expressed in pA/μm², conductances in nS/μm², the capacitance of the cells is normalized to 1 pF/μm². Parameter values can be found in the appendix 2 figures 39 and 40 show the steady state activation and inactivation variables of I_{Ca} and I_T.

5.2.2
Spontaneous Activity

Corresponding with Terman et al. (2002), the reproduction of the model STN neuron shows spontaneous activity with a firing rate of approximately 3 Hz (Fig. 38). Zooming in on a single action potential results in Fig. 41.

The duration of a single action potential is approximately 2.5 ms, which is clearly longer than the duration reported by Nakanishi et al. (1987); 1 ms. The resting membrane potential is roughly −57 mV, with membrane potentials varying from −70 to

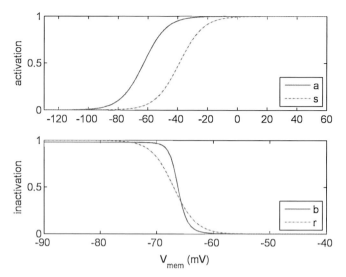

Fig. 39 *Upper graph*: Steady-state voltage-dependent activation variable of the high-threshold Ca^{2+} current, s_∞, and the activation variable of the low-threshold T-type Ca^{2+} current, a_∞. *Lower graph*: inactivation b_∞ which is a function of r. The steady state of the inactivation variable r_∞ is included to point out the difference between the two

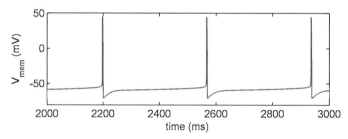

Fig. 40 Spontaneous activity of the modelled STN neuron model at around 3 Hz. (Reproduced from Terman et al. 2002)

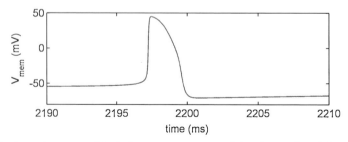

Fig. 41 Single action potential during spontaneous activity. (Reproduced from Terman et al. 2002)

+50 mV during action potentials. The fast hyperpolarization phase can be discerned; however, no slow afterhyperpolarization and clear slow-ramp depolarization, as observed by Nakanishi et al. (1987), are present during spontaneous activity.

The steady-state I–V curves shown in Fig. 42 shows the TTX-sensitive sodium current with negative slope conductance comparable to the curves experimentally observed by Bevan and Wilson (1999) (see Fig. 5, Sect. 3.1).

With depolarizing input currents, the firing rate increases up to roughly 200 Hz in a near-linear curve (Fig. 43). This curve suggests that with increasing current spike frequency increases even further.

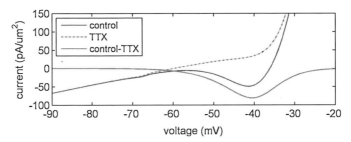

Fig. 42 Negative slope conductance resulting from a TTX-sensitive sodium current

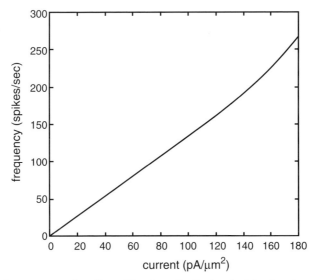

Fig. 43 Spike frequency as a function of the strength of a depolarizing input current. (Source: Terman et al. 2002)

5.2.3
Rebound Bursts

At the break of hyperpolarizing input currents, the STN model generates rebound bursts. Figure 1E (see Fig. 44) and 1F of the paper of Terman et al. (2002) could be accurately reproduced. According to Terman, increasing the length or the amplitude of the current pulse results in an enhanced rebound. However, the maximum rebound burst length was about 200 ms.

The currents involved in the generation of rebound bursting are investigated by applying a hyperpolarizing current pulse while blocking Na⁺ currents to suppress spontaneous spiking. Figure 45 shows the resulting membrane potential and the activation and inactivation variables.

During the hyperpolarizing pulse, the T-type current is slowly deinactivated ($b_\infty \to 1$). At the break of the current pulse, this current is activated ($a_\infty \to 1$) but also slowly inactivated ($b_\infty \to 0$).

Terman et al. (2002) do not mention plateau potentials and also do not indicate that they have tested the response of the STN neuron model when the membrane potential was kept at a hyperpolarized state. It was found that the STN neuron model of Otsuka et al. (2004) (see previous section). This was investigated by applying a constant inhibitory input current to bring the membrane potential down to the range that was found to be required for plateau potentials to be generated (see Sect. 3.2). At mild hyperpolarized states of the membrane by the constant input of a hyperpolarizing current, rebound bursts could still be generated (Fig. 46).

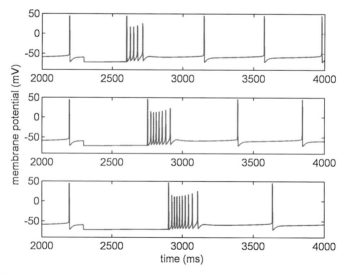

Fig. 44 Rebound responses to a hyperpolarizing input current of −25 pA/μm² with a duration of 300, 450, and 600 ms from top to bottom, respectively. (Reproduced from Terman et al. 2002)

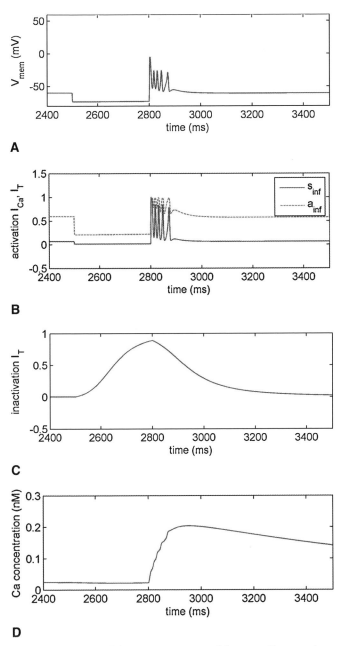

Fig. 45 A Membrane potential of the STN neuron model responding to a hyperpolarizing input of −30 pA/μm² with a pulse width of 300 ms, starting at $t=2{,}500$ ms while Na⁺ currents are blocked, i.e. $g_{Na}=0$. B Instantaneous activation variables s_∞ and a_∞, of current I_{Ca} and I_T, respectively. C Instantaneous inactivation variable b_∞ of current I_T. D Internal calcium concentration

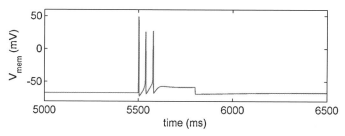

Fig. 46 Responses to a depolarizing input current of 20 pA/μm² with a duration of 300 ms while the membrane was kept at a hyperpolarized state resulting from a constant input of −20 pA/μm²

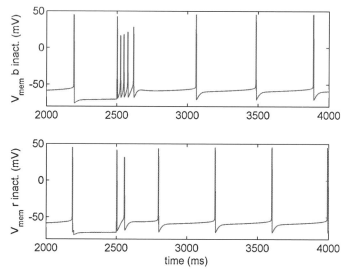

Fig. 47 Rebound response following a hyperpolarizing input pulse of −25 pA/μm² for 300 ms while b (*upper graph*) and r (*lower graph*) describe the inactivation of the T-type current

However, for larger hyperpolarized membrane potentials, the neuron was completely silenced.

The way the T-type Ca²⁺ channel was described seems rather artificial. In testing the effect of using b instead of r for the inactivation of this channel, the results as shown in Figs. 47 and 48 were obtained.

5.3
Comparison of the Otsuka Model with the Terman/Rubin Model

The firing behaviour at resting membrane potentials, the firing response to current injection, as well as a long-lasting rebound potential in STN neurons were simulated in both models.

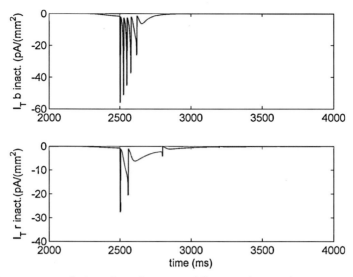

Fig. 48 T-type current during rebound response following a hyperpolarizing input pulse of −25 pA/μm² for 300 ms, while b (*upper graph*) and r (*lower graph*) describe the inactivation of the T-type current

As can be observed from the equations and parameter values, dimensions and scaling are completely different. For example, Otsuka et al. (2004) describe currents in μA/cm² (=10^{-2} A/m²) while Terman et al. (2002) express currents in pA/μm² (=A/m²).

Both models show spontaneous activity that may be explained by a negative slope conductance in the range associated with the resting phase (at around −58 mV), as observed in both models in their steady-state I–V curves. Terman's model gives rise to spontaneous activity at nearly 3 Hz, while Otsuka's model arrives at nearly 5 Hz. These firing rates are at the lower limit of those observed in experimental studies in which firing rates of 5–40 spikes/s have been found. Rubin and Terman (2004) apply an additional constant current of 25 pA/μm² in order to increase the firing rate.

The shape of the action potential is more realistic in the model of Otsuka et al. (2004); however, in both situations, peak duration is longer than those observed experimentally.

With increasing magnitude of depolarizing current pulses, the firing frequency of the STN neurons increases (in a near linear manner) in both models; however, the model of Otsuka et al. (2004) found a maximum firing frequency of 100 Hz, which is far below the experimentally observed maximum of approximately 500 Hz.

Both models demonstrate bursting activity at the break of hyperpolarizing inputs. Otsuka et al. (2004) focussed on the voltage-dependent generation of plateau potentials and rebound potentials by investigating the effect of bringing the membrane into a hyperpolarized state. They included L-type currents as playing an essential role in the voltage-dependent generation of plateau potentials and long-lasting rebounds, while Terman et al. (2002) simulated long-lasting rebound potentials with T-type Ca^{2+} currents responsible for the underlying dynamics.

STN neurons can be divided into non-plateau-generating and plateau-generating neurons according to Beurrier et al. (1999). The differences in ion channel dynamics regarding these two types have not been investigated. Since there is no clear definition of a plateau potential, it may be difficult to tune the model parameters to these two types. Based on experimental observations, Otsuka et al. (2001) reported that non-plateau-generating neurons are also able to generate rebound potentials after termination of hyperpolarizing current injection. These rebound potentials, however, were terminated within 100 ms and were mediated by T-type currents. The model of Otsuka et al. (2004) does show the ability to generate rebound potentials without generating a plateau potential; however, the duration of this burst is comparable to the burst duration generated in combination with a plateau potential; compare Figs. 32 and 33.

Terman and Rubin (Terman et al. 2002; Rubin and Terman 2004) expect the ability to generate bursting activity to represent the parkinsonian condition. No additional hyperpolarization was applied. In addition, no clear long-lasting depolarization, i.e. a plateau potential, was present during this bursting activity (see Fig. 44), and burst duration was limited to roughly 200 ms. A decrease in the firing rate during rebound bursts as well as the duration of rebound responses in relation to the amplitude and duration of the input current pulse are more realistically simulated by the model of Otsuka et al. (2004). It may therefore be concluded that the Otsuka model shows the best comparison to the results obtained from the experimental situation.

Different membrane channels in STN neurons normally also represent input from different neurotransmitter systems. The connective architecture of the STN neurons (see Part I and Terman et al. 2002) and its consequences are hardly represented in these single-compartment models.

While the model of Rubin and Terman (2004) introduces DBS in STN, the response of the STN neuron model is not discussed. It may be worthwhile to simulate the response of a single STN neuron to stimulation at different frequencies. In the network model of Rubin et al. (2004), described in Sect. 5.6 (this volume), STN-DBS is introduced by applying a stimulus current to the STN according to the following equation

$$I_{DBS} = i_{DBS} H(\sin(2\pi t / \rho_{DBS}))[1 - H(\sin(2\pi(t + \delta_{DBS}) / \rho_{DBS}))]$$

with H the Heaviside step function ($H(x)=0$ for $x<0$ for $x<0$, and $H(x)=1$ for $x \geq 0$), ρ_{DBS} is the period, i_{DBS} is the amplitude of the signal, and δ_{DBS} is the duration of the positive phase.

Using a pulse width of 200 μs comparable to clinically used values, the effect of stimulation is investigated in both STN neuron models for a stimulation frequency of 50, 130 and 180 Hz; results are shown in Figs. 49 and 50 for the model of Otsuka and Terman, respectively (with a maximum time step of 10 μs). In both models, an

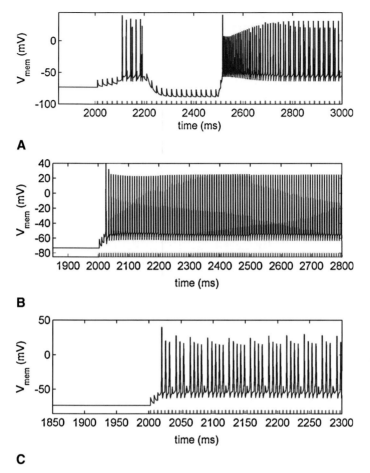

Fig. 49 Stimulation of the STN model neuron of Otsuka et al. 2004 by applying pulses of 250 µA/cm^2 with a duration 200 µs and a frequency of **A** 50 Hz; **B** 130 Hz; **C** 180 Hz. In all situations, the membrane is hyperpolarized by a constant input of −5 pA/µm^2, while during 300 ms (from $t=2{,}200$ ms to $t=2{,}500$ ms) an inhibitory input of −5 pA/µm^2 is applied

additional input, e.g. representing an input from GPe, is applied for 300 ms, starting at $t=2{,}200$ ms. While in Otsuka's model an additional constant hyperpolarization is applied to transform the model into the parkinsonian situation, Terman did not decrease the membrane potential of STN neurons in the network model as a representation of PD.

While the amplitude of the stimulation-driven action potentials are largely decreased in the model of Terman et al. (2002), both models show that at a frequency of 130 Hz, comparable to the clinically used frequency, STN activity

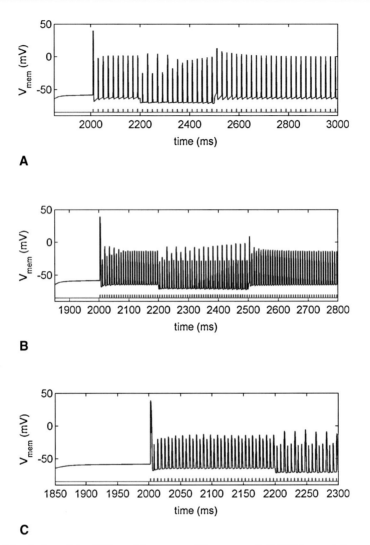

Fig. 50 Stimulation of the STN model neuron of Terman et al. (2002) by applying pulses of 250 µA/cm² with a duration 200 µs and a frequency of **A** 50 Hz; **B** 130 Hz; **C** 180 Hz. In all situations, the membrane is at resting potential, while during 300 ms (from $t = 2{,}200$ ms to $t = 2{,}500$ ms) an inhibitory input of -20 µA/cm² is applied

is completely driven by stimulation, with Otsuka's model showing the least response to other incoming signals. The inability to respond to additional inputs when stimulated may be the mechanism of the clinical effect of DBS in comparison to the results from the network model of Rubin and Terman (2004) that considers the hypothesis that DBS works by replacing pathologically rhythmic basal ganglia output with tonic, high-frequency firing (see Sect. 5.6.2).

5.4
The Multi-compartment STN Model of Gillies and Willshaw

"The extent of neural activation generated by extracellular stimulation depends on the stimulation parameters, electrode and tissue properties, and the position and orientation of neural elements with respect to the electrode" (Miocinovic et al. 2006).

Moreover, STN stimulation could very well also stimulate the fibres of field H_2 and the zona incerta and the corticospinal tract (see also Part I; Miocinovic et al. 2006). It is inevitably that the Bauplan of the STN and its surroundings therefore require a three-dimensional approach.

Multi-compartment models for the electrical behaviour of an STN neuron can be based on morphological parameters such as soma size, dendritic diameters and spatial dendritic configuration in combination with their electrical properties. Experimental data of the exact ion channel dynamics of the human STN neuron is scarce. These gaps in knowledge are filled in by using the parameters obtained from experimental animal studies (in the case below, channel dynamics are taken from thalamus and cortex).

5.4.1
Membrane Dynamics

A computational multi-compartment model of the rat subthalamic projection neuron has been constructed by Gillies and Willshaw (2006). The morphological specification of the rat STN projection neuron has been taken from Afsharpour et al. (1985) and Kita et al. (1983) (see also Part I, Sect. 2.1). The morphology of the STN neuron was implemented in the model and the arrangement and length of the compartments in the model are based on measured data (Fig. 51). Four

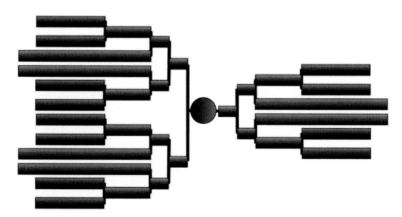

Fig. 51 STN cell morphology as specified by Gillies and Willshaw. (Adapted from Gillies and Willshaw 2006)

parameters were used to model the distribution and density of ion channels: (1) the channel density at the soma; (2) the overall density across all the dendritic trees; (3) the amount of density that is uniform across the dendritic trees; and (4) specification of the linear distribution ranging between −1 (maximally distal) and 1 (maximally proximal).

Passive membrane properties are described according to:

$$C_{m,i}\frac{dv_i}{dt} = \frac{v_{i-1} - v_i}{r_{i-1,i}} - \frac{v_i - v_{i+1}}{r_{i,j+1}} - I_{ion,i}$$

with for each compartment i, $C_{m,i}$ the membrane capacitance, v_i the membrane potential and $r_{i-1,i}$ and $r_{i,i+1}$ the axial resistance between compartment i and its previous and following compartments. $I_{ion,i}$ is the combination of the ionic currents in the compartment, which in case of a passive membrane consists of a leak current only.

Active membrane properties are described as follows:

$$I_{ion,i} = I_{Na,i} + I_{NaP,i} + I_{KDR,i} + I_{KV31,i} + I_{sKCa,i} + I_{h,i} + I_{CaT,i} + I_{CaL,i} + I_{CaN,i} + I_{leak,i}$$

The following is a brief description of the channels supporting the ion current (details can be found in Gillies and Willshaw 2006):

I_{Na} Fast-acting Na$^+$ channel
I_{NaP} Persistent Na$^+$ channel
I_{KDR} Delayed rectifier K$^+$ channel
I_{Kv31} Fast rectifier K$^+$ channel
I_{sKCa} Small conductance Ca^{2+}-activated K$^+$ channel
I_h Hyperpolarization-activated cation channel
I_{CaT} Low-voltage-activated T-type Ca^{2+} channel
I_{CaL} High-voltage-activated L-type Ca^{2+} channel
I_{CaN} High-voltage-activated N-type Ca^{2+} channel
I_{leak} Leak current

5.4.2
Activity Patterns

One of the main goals of Gillies and Willshaw (2006) was to construct a model that exhibits many of the well-known and characteristic features of STN neurons employing a reduced or minimized parameter set. The set of characteristics include (see Fig. 52):

1. The action potential and hyperpolarization properties characterized and recorded by Beurrier et al. (1999).
2. The in vitro resting firing patterns recorded at different temperatures by Bevan and Wilson (1999) and Bevan et al. (2002).

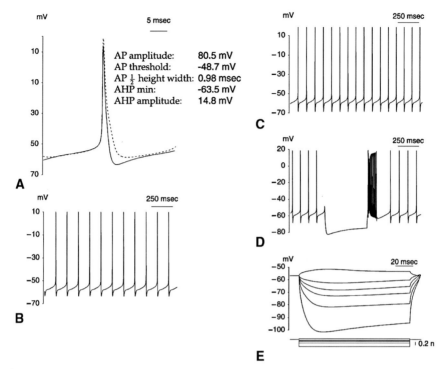

Fig. 52 Simulations of the characteristic features of STN cells. **A** Action potential (feature 1); **B, C** spontaneous activity for a temperature of 25°C and 35°C, respectively (feature 2); **D** rebound burst at the break of a hyperpolarizing input (feature 3); **E** passive properties (feature 4). (Source: Gillies and Willshaw 2006)

3. The hyperpolarization response with a sag and the presence of a rebound response characterized by Bevan et al. (2002).
4. The passive properties according recorded transients observed in in vitro intracellular experiments of, for example, Nakanishi et al. (1987).
5. The repetitive burst firing during a hyperpolarizing stimulus and simulated apamin protocol in comparison with observations from, for example, Beurrier et al. (1999).

Three channel types were found to play an important role in distinguishing the experimentally observed characteristic activity patterns:

- A high-voltage-activated T-type Ca^{2+} channel
- A low-voltage- activated L-type Ca^{2+} channel
- A small conductance Ca^{2+}-activated K^+ channel

although nine channel types were implemented (see above, and Gillies and Willshaw 2006 and references therein). In accordance with several experimental studies, the T-type Ca^{2+} channel was found to be necessary as a trigger of many of the behaviours.

The interactions of these channels demonstrated:

- Short and long posthyperpolarization rebound responses.
- Low-frequency rhythmic bursting (<1 Hz).
- High-frequency rhythmic bursting (4–7 Hz).
- Slow action and depolarizing potentials.

The exact values of the key passive properties of STN neurons are difficult to determine according to Gillies and Willshaw (2006). Moreover, the width of the action potential is larger than recorded experimentally, presumably because of the non-optimal parameter selection or the channel kinetic descriptions. The linear distribution of channel densities gives a distal bias on the location of the mechanisms that produce the plateau potentials.

In fact, "the density and the distribution of channels in the model are uncertain parameters in these simulations. Experimental techniques do not yet exist to give a detailed distribution and density information for all ionic channels in a given cell" (Heldoorn et al. 2003; for STN, see Part I, Sects. 2.3.1–2.3.5). The final stage of constructing a complex realistic model involves tuning model parameters to replicate well-described properties of the neuron in question. The description provided by Gillies and Willshaw (2006) gives a clear overview of the tuning and how the additional questions are explored.

Difficulties in fitting and tuning the model parameters result from:

- The changing environment during procedures for brain slice preparation and cell dissociation may result in cell shrinkage and changes in the electrical properties of the cells.
- Different culturing media.
- Different temperatures during the experiments, which makes it difficult to compare results.

5.5
Intra-nuclear Network Models

To the best of our knowledge, intra-nuclear network models of the STN are not found in the literature. Such models can be made using an active soma, initial axon segments and dendritic trees in comparison to the model described in the previous section. The neurons can be connected by collateral or synaptic connections or electrotonic connections, and even inter-neurons can be modelled. Such a model was presented for Onuf's nucleus (Heldoorn et al. 2001a, b) only containing electrotonic coupling between the neurons. This type of intra-nuclear network model is missing for the STN. This is the all the more disappointing since the presence of collateral input from STN neurons to other STN neurons is debated (see Part I).

5.6
Inter-nuclear Network Models

Based on experimental studies, it has been proposed that under normal physiological conditions STN and GPe neurons do not generate correlated spontaneous rhythmic activity or even show internally correlated firing. In contrast, correlated oscillatory activity patterns in GPe and STN neurons are suggested to underlie the generation of the symptoms of PD. In organotypic cultures, it was found that this synchronized activity is caused by the interaction between STN and GPe rather than being driven by an external source (Plenz and Kitai 1999).

Dopamine normally depolarizes STN membrane potentials. Therefore, in normal subjects GP inhibitory inputs would not hyperpolarize STN neurons enough for plateau potentials to occur. Dopamine depletion, on the contrary, may result in STN membrane hyperpolarization beyond the level normally induced by GP inhibitory inputs, making the STN neurons susceptible to generating plateau potentials. STN neurons project to GPe cells via glutamatergic receptors, while GPe cells inhibit each other via recurrent axon collaterals. This closed loop in the indirect pathway of the basal ganglia may form the basis of a central pattern generator that may be responsible for the oscillatory activity patterns, as found under parkinsonian conditions; GPe neurons that are excited by STN neurons in turn inhibit STN cells, after which a rebound burst may be generated, which again excites GPe neurons (Plenz and Kitai 1999).

Terman et al. (2002) investigated the conditions that are required for this oscillatory behaviour in terms of the arrangements and strengths of synaptic connections within GPe and between STN and GPe populations. A summary of their results is given here.

5.6.1
GPe-STN Network

Terman et al. (2002) explore the dynamic interactions of the network of the subthalamic nucleus and the external segment of the globus pallidus by conductance-based computational models. The STN neuron model has already been described in Sect. 5.2 (this volume). Similarly, the membrane potential v of the single compartment model of the GPe is described by:

$$C_m \frac{dv}{dt} = -I_{Na} - I_K - I_{Ca} - I_T - I_{AHP} - I_{leak} + I_{app} - I_{GPe \to GPe} - I_{STN \to GPe}$$

in which the incorporated ionic currents are similar to the equations used for the STN cell, except for the T-type current, for which a simpler equation is used:

$$I_T = g_T a_\infty^3(v) r (v - v_{Ca})$$

with r satisfying a first-order differential equation (dw/dt) with τ_r constant.

I_{app} is a constant hyperpolarizing current representing the input from striatum, which is assumed to be a common input to all GPe cells. Without any inputs, GPe cells are spontaneously active; from Fig. 2 of Terman et al. (2002), a spiking frequency of about 29 Hz is observed.

$I_{GPe \to GPe}$ represents the synaptic input from recurrent connections in GPe:

$$I_{GPe \to GPe} = g_{GPe \to GPe}(V - V_{GPe \to GPe}) \sum_{i=1}^{n} s_i$$

while the synaptic input from STN to GPe is described by:

$$I_{STN \to GPe} = g_{STN \to GPe}(v - v_{STN \to GPe}) \sum_{i=1}^{n} s_i$$

Similarly, the STN receives input from GPe according to:

$$I_{GPe \to STN} = g_{GPe \to STN}(v - v_{GPe \to STN}) \sum_{i=1}^{n} s_i$$

$g_{GPe \to GPe}$, $g_{GPe \to STN}$, and $g_{STN \to GPe}$ are the synaptic conductance from GPe to GPe, from GPe to STN, and from STN to GPe, respectively. The (GABAergic) synaptic coupling from GPe to STN is inhibitory with a reversal potential of −85 mV, while the reversal potential for the inhibitory recurrent connections in GPe is −100 mV. The (glutamatergic) synaptic coupling from STN to GPe is excitatory with a reversal potential of 0 mV. The summation is taken over the presynaptic neurons according to synaptic variables described as: $\frac{ds_i}{dt} = \alpha H_\infty(vg_i - \theta_g)[1 - s_i] - \beta s_i$ with vg_i the membrane potential of the presynaptic neuron, nr. i, and the function $H_\infty(v) = \frac{1}{1 + \exp[-(v - \theta_g^H)/\sigma_g^H]}$. Parameter values can be found in the appendix 2.

5.6.1.1
Network Architectures

Simulation results show that the cellular properties of STN and GPe neurons can give rise to a variety of rhythmic or irregular self-sustained firing patterns, depending on the arrangement of connections among and within the nuclei and change in the connection strengths. Three prototype networks are proposed, with the number of cells for each neuron type ranging between 8 and 20:

1. Random, sparsely connected architecture: each GPe cell sends inhibitory input to a small proportion of the STN, and each STN cell sends excitatory input to a small proportion of the GPe.
   Firing is irregular and weakly correlated with the activity of other cells when there is a weak connection from STN to GPe or a strong inhibition from GPe to GPe.

Strong excitatory connections from STN to GPe and weak inter-GPe connections give rise to continuous uncorrelated activity. Intermediate connection strengths give rise to episodic spiking with episodes lasting for approximately 300 ms and increasing duration, with increasing connection strength from STN to GPe ($g_{STN \to GPe}$). Episodes are dependent on calcium uptake in both cells; the internal calcium concentration increases with spiking. The end of an episode is determined by I_{AHP} of GPe cells. The time between episodes is primarily determined by I_{AHP} in STN cells.

2. Structured, sparsely connected architecture: each GPe cell inhibits its two immediate GPe neighbours and two STN cells skipping the three nearest neighbours. Because of this off-centre architecture, lateral inhibition among GPe cells that have overlapping projections in the STN is created. Each STN cell excites only the nearest GPe cell.

Varied network dynamics result from this architecture, with most patterns featuring clustering with subsets of highly correlated neurons. Activity may switch from one cluster to another. Cluster alternation is mainly driven by persistent inhibition and the resulting removal of inactivation of rebound currents. The rate of deinactivation of I_T in inactive STN cells relative to the level of inhibition determines the duration of the active episodes; in the study presented, the switching of clusters occurred at a rate of approximately 5 Hz.

3. Structured, tightly connected architecture: each GPe cell contacts the five nearest neighbours in STN and all other GPe cells. Each STN cell contacts the three

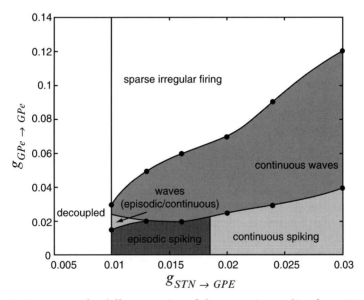

Fig. 53 Activity patterns for different setting of the synaptic coupling from GPe to GPe ($g_{GPe \to GPe}$) and from STN to GPe ($g_{STN \to GPe}$); coupling from GPe to STN is set at 1 nS/μm² and striatal input is set at −1.2 pA/μm². (Source: Terman et al. 2002)

nearest neighbours in GPe. Depending on the strengths of connectivity, different patterns may occur in the network, as shown in Fig. 53. Simulating several related architectures revealed that the existence of propagating waves was a very robust activity pattern. Wave propagation requires a structured architecture so that active STN and GPe cells spread activity to cells ahead of the leading edge. In randomly connected networks, this type of activity will not occur.

In case of a weak striatal input, GPe cells are allowed to fire tonically, leading to tonic inhibitory input to STN. This input may completely suppress STN activity, and thus in this situation GPe is less sensitive to STN input. In addition, there is strong GABA-A synaptic inhibition among GPe neurons, making their output to the STN asynchronous, and effectively weakening the synaptic interactions between GPe and STN.

Terman et al. (2002) propose that after dopaminergic denervation, an increased level of inhibition from the striatum to GPe is combined with the release of enkephalin and dynorphin, which acts presynaptically to weaken the collateral connections among GPe cells. As confirmed by the simulation results, increasing the (constant) striatal input I_{app} into GPe neurons, and a weakened intra-pallidal (inhibitory) coupling, i.e. a decreased $g_{GPe \to GPe}$ may shift the network into an oscillatory (PD) mode. The increased STN activity results in an increased inhibitory output from basal ganglia to thalamus, causing the hypokinetic symptoms associated with PD.

Since network architecture largely determines activity patterns within the basal ganglia, as pointed out by the authors, it is crucial to know the spatial extent of the recurrent collateral connections among GPe neurons, whether they are spatially organized or diffuse, and whether the two nuclei project on each other reciprocally or out of register. For example, the network should be able to support spatially organized activity with a wide range of phase relationships, as observed in MPTP-treated monkeys (Raz et al. 2000). In addition, Parkinson's tremor also shows a variety of phases in different parts of the body (Hurtado et al. 2000; Ben-Paz et al. 2001), suggesting multiple oscillatory subnetworks to be present in the STN-GPe network.

Besides this lack of knowledge concerning network connections, the model has some additional shortcomings:

- Incorporated properties of STN as well as GPe cells were based on experimental data from slice preparations, which may somehow differ from in vivo situations.
- No anatomical information is included by Terman et al. (2002); for example, the effects of synaptic contacts close to the soma are stronger than those at a larger distance; how long are the dendrites and axons of the STN and GPe neurons and to what extent are these different types of neurons connected to each other? These parameters are included within $g_{STN \to GPe}$ and $g_{GPe \to GPe}$, which represent the connectivity between STN-GPe and the inter-connectivity of GPe, respectively (see also Part I).
- Neurotransmitter/receptor types are not included (see also Part I).
- No inter-connectivity between STN cells is included, while the model of Gillies and Willshaw (2004) indicated that there is at least 3% connectivity among STN

neurons. These axon collaterals, however, are denied by Kitai et al. (1983) and Hammond and Yelnik (1983) (see also Part I).
- As indicated by the authors themselves, the constant striatal input into GPe is not a realistic representation. It was used as an approximation to the overall effect of the inhibitory effect of the striatum on GPe.

5.6.2
GPe-STN-GPi-Thalamus Network

As indicated, experimental data reveal that the output nuclei of the basal ganglia (GPi) become overactive in PD, increasing the level of inhibition in thalamus. High-frequency stimulation was found to increase activity even more in stimulated areas, from which the beneficial effect is hard to grasp. By extending the GPe-STN model described in the previous section with GPi and thalamic cells (Fig. 54), Rubin and Terman (2004) show that a tonic firing of GPi as oppressed by DBS may restore thalamic relay capabilities.

STN and GPe neuron types are described using equations similar to the ones described in Sect. 5.2.1 and 5.6.1 (this volume), respectively. GPi neurons were modelled exactly as the GPe neurons; however, in order to match firing frequencies with in vivo data instead of in vitro data, additional depolarizing constant inputs were applied to STN (25 pA/μm²), GPe (2 pA/μm²), and GPi (3 pA/μm²).

In the network two thalamic neurons are modelled. They are described according to:

$$C_m \frac{dv}{dt} = -I_{Na} - I_K - I_{Ca} - I_T - I_{leak} - I_{GPi \to Th} + I_{SM}$$

These cells are supposed to act as a relay station of incoming sensorimotor signals I_{SM}, which are represented by periodic step functions.

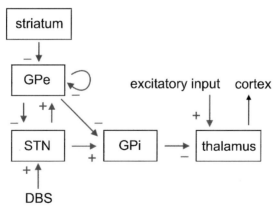

Fig. 54 Part of the basal ganglia network as modelled by Rubin and Terman. (Adapted from Rubin and Terman 2004)

$$I_{SM} = I_{SM} H\left(\text{Sin}(2\pi t / \rho_{SM})\right)\left[1 - H\left(\text{Sin}(2\pi(t + \delta_{SM}) / \rho_{SM})\right)\right]$$

with H the Heaviside step function, as defined in Sect. 5.3, ρ_{SM} the period, i_{SM} the amplitude of the signal, and δ_{SM} the duration of the positive phase. They also receive input from the basal ganglia via GPi, $I_{GPi \to Th}$. Further model details can be found in Rubin and Terman (2004).

Thalamic cells are not spontaneously active but respond to depolarizing inputs with continuous spiking. At the break of hyperpolarizing inputs, rebound bursts are generated comparable to the bursting activity that correlates with tremor frequency, as observed in PD. Larger hyperpolarizing currents lead to stronger rebounds that eventually may disrupt the relay of sensorimotor input to the cortex.

Testing the relaying capabilities of the thalamic cell by assuming constant inputs from GPi, it was observed that for inhibitory inputs the response is frequency-dependent. A hyperpolarizing input will deinactivate the T-type Ca^{2+} current in the thalamic cell. However, for fast-input pulses this current does not have enough time to deinactivate sufficiently, resulting in one brief action potential instead of bursting activity.

The simulated network consists of 16 STN neurons, 16 GPe neurons, 16 GPi neurons and two thalamic neurons. Each STN received inhibitory input from two GPe neurons. Each GPe neuron received excitatory input from three STN neurons and inhibitory input from two GPe neurons. Each GPi neuron received excitatory input from one STN neuron. Each thalamic neuron received inhibitory input from eight GPi neurons. Whether the connections involve nearest neighbours, off-centre or randomly selected neurons was not indicated.

The network can produce both irregular asynchronous activity and synchronous tremor-like activity. Irregular asynchronous activity is regarded as the normal state. With increased striatal input to GPe (I_{app}) and decreased inter-pallidal inhibition ($g_{GPe \to GPe}$) the network switches to the synchronous activity representing the parkinsonian state, according to experimental results. With a pulsed input into all STN cells, the STN-DBS state is simulated.

5.6.2.1
Normal State

In the normal state, the thalamus receives uncorrelated, irregular input from GPi, which does not disrupt the relaying capability of the thalamus in responding faithfully to the excitatory sensorimotor input.

5.6.2.2
Parkinsonian State

In the parkinsonian state, the GPi shows phases of intense activity resulting from phasic GPe bursts. In this case, the relaying function of the thalamus is disturbed. Sensorimotor pulses are missed or the TC cells start to burst.

5.6.2.3
STN-DBS State

The DBS signal driving the STN is modelled similarly to the pulsed signal of the sensorimotor signal (see also Sect. 5.3). Stimulation in STN results in tonic firing of GPi, resulting in strong but tonic inhibition in thalamic cells. Reducing the oscillatory nature of the inhibition of thalamic cells as in PD may be the key mechanism of DBS efficacy. Stimulation with an amplitude of 200 pA/μm^2, a stimulation period of 6 ms (i.e. roughly 167 Hz) and a pulse duration of 0.6 ms, completely restores the thalamic ability to relay the sensorimotor signal.

The introduction of stochastic noise in the time interval between sensorimotor inputs, ranging from 35 to 80 ms, was found not to influence the responsiveness of the TC cell for the three different states. Under these noisy conditions, the performance of DBS declines if the intensity of DBS becomes too strong or the frequency becomes too high. It is reported that for sufficiently large amplitudes, DBS improves the thalamic relaying capability over a wide range of stimulation pulse durations and frequencies that are qualitatively comparable to results from clinical evaluation. In addition, at low frequencies DBS efficacy is significantly reduced both in clinical practice and simulation.

Rubin and Terman (2004) investigate the responsiveness of a thalamic cell to combinations of inhibitory GPi and excitatory inputs using phase plane methods. A detailed description of this analysis is beyond the scope of this monograph; however, the methodology provides a powerful tool in analysing the response of a single cells for different input signals without knowing the exact source of the driving force.

Rubin and Terman (2004) conclude their paper with a description of the shortcomings of their model:

- Cells are represented by single-compartment models, and thus no dendrite and axon properties are included.
- The source of the excitatory sensorimotor input into the TC cell has not been specified.
- Thalamic reticular cells (RE) are not included: pathological conditions may trigger bursts in the inhibitory thalamic reticular cells via GPi or GPe, which would in turn induce burst activity in TC cells throughout the thalamus.
- It is assumed that the role of the GPi is simply to relay input from STN; the modulatory function of the basal ganglia is not considered.
- Certain connections within the basal ganglia are neglected, such as those from striatum to GPe.
- Electrode positioning as well as the extent and effectiveness (enhancement or suppression of neuronal activity within a particular structure and influence of geometry and orientation of neurons) of the stimulation field within the STN are not considered.

A number of shortcomings or questions may be added to this list:

1. Representing thalamic cells as simple relay stations may be an oversimplification of the function of these cells, while using only two thalamic cells may not give an accurate average result.
2. The application of a constant current as a representation of striatal input into GPi may be inadequate.
3. Application of constant currents to increase the firing rates of different cell types according to in vivo recordings may obscure the underlying dynamics of the neurons.
4. The choice of the connectivity scheme is not well motivated (see Part I, and Sect. 5.6.1 of this volume.

No attempt has been made to reproduce the results of the simulated network models described by Terman et al. (2002) and Rubin and Terman (2004). Pascual et al. (2006) were able to partially reproduce the results of Rubin and Terman (2004) and questioned the usefulness of these models and their contribution to increasing our understanding of the network behaviour under normal, parkinsonian and DBS conditions. They concluded that modelling basal ganglia network activity is complicated by the following considerations:

1. The basal ganglia circuitry is extremely complex and therefore it is difficult to grasp all the necessary details.
2. It is difficult to quantify the stimulation field in time and space as well as its effect on the different neuronal elements.
3. It is difficult to record behavioural data and neuronal activity simultaneously in the different locations necessary to validate models describing DBS.
4. A wide variety of mathematical formalisms and scales between and within models exist.
5. A large number of model parameter values need to be specified, which may easily induce errors.

While Rubin and Terman (2004) tested the robustness of the model by varying the spike interval of the sensorimotor input, Pascual et al. (2006) tested model robustness by simulating the model under exactly the same conditions with two parameters of the model modified by 5%; they came to the conclusion that the model lacked robustness for small changes: one of the thalamic cells switched from a perfect to a completely unreliable relay function. However, they do not well motivate the choice for the two parameters they modify (θ_r and σ_{rt}, both related to the inactivation variable r of the T-type Ca^{2+} current of the thalamic cell; decreasing θ_{rt} shifts the inactivation to lower membrane potentials). It may even be debated whether it would have been more correct to test the robustness of a model by varying the parameters one at a time.

Pascual et al. (2006) came to the conclusion that modelling the effect of DBS in PD should be approached from a different angle. Unfortunately, no new ideas or methods are proposed.

6
Comparison of Part I and Part II

To understand the function of the human STN in normal and disease situations, especially PD, modelling, human neuroanatomical degeneration studies and human receptor localizations are important. Several topics remain undecided as to an adequate anatomical and physiological approach that contributes to application in DBS.

6.1
Recurrent STN Axons

In 1983, Kitai et al. described the presence of recurrent axon collaterals that are present within the rat STN. The amount is hard to ignore; it concerned 50% of the STN axons. No indication of the presence of these recurrent collaterals is described for all other species considered in this monograph (see Part I, Sect. 2.1). Recurrent glutamatergic excitatory connections within the STN were ruled out by the physiological study of Bevan and Wilson (1999), also in the rat (see Sect. 3.1.1, this volume). These recurrent axon collaterals could be important for the spontaneous activity present in the STN. Therefore, fluorescent dye injections in the STN neurons of several species are a necessary prerequisite to knowing whether such recurrent axons are present. Moreover, these fluorescent dye studies (e.g. DiI) can also be carried out in human post-mortem material.

6.2
Inter-neurons in the STN

In the rat, inter-neurons are said not to be present in the STN. All other studied species in this monograph presumably have inter-neurons (Part I, Sect. 2.1). Inter-neurons are mostly of the inhibitory type and use GABA. The presence of GABA-A and -B receptors (Part I, Sect. 2.3.4.5), therefore, not only is an indication for the functionality of GABAergic projections towards the STN (see Boyes and Bolam 2007), but presumably also for the activity of the STN inter-neurons. Since the rat is used in most physiological studies but lacks inter-neurons and is an open nucleus, while all other STN nuclei are closed and contain inter-neurons, one should reconsider whether the rat results can be translated to humans.

6.3
Fibre Tracts around and in the STN

Several authors consider the possibility that antidromic stimulation via the cerebral peduncle or other systems could be responsible for the effect of DBS (Sect. 2.3.2, volume). Although in most species the connection between cortex and STN is established, there is serious doubt that this also holds for humans (Part I, Sect. 5.2.1). The diameter of the axons needed to explain the antidromic velocity seems not (or only in a very restricted manner) to be present in the human corticofugal system. An alternative pathway has been proposed that could reach the STN and the pedunculopontine nucleus, connecting the STN activity with the initiation of movement (Part I, Sect. 5.2.4).

6.4
Ca^{2+} Receptors

Series of physiological results on Ca^{2+} receptors are only partially supported by localization of the receptor at the ultrastructural level (see Part I, Sect. 2.3.5). Localization and function coincides for the L-type; however in situ hybridization indicates the presence of the P-type, while physiology refutes its functionality in the STN. For modelling (see Sect. 5, this volume), an absolute condition is to have an overview of the receptors in the STN with their ultrastructural localization and functionality. This is especially true for the Ca^{2+} receptors, since they are the keys to plateau potentials and synchronization of STN neuronal activity (Sect 3.2 and 3.3, this volume)

6.5
Three-Dimensional Modelling

The somatotopy present for several systems in the STN (see Part I, Sects. 4 and 5) requires 3D modelling. Moreover, orthodromic or antidromic stimulation of systems by DBS in the STN also supports 3D modelling. The position of the neurons and their appendices, like axons, dendrites, synaptic contacts and passing tracts towards the electrodes, determines its effects and can only be modelled for the human situation in a 3D configuration (see Sect. 5.4, this volume, and Part I, Sect. 5). Within such a 3D configuration, the effects of inter-neurons will be studied more effectively, at least better than in a 2D set-up.

6.6
Types of Projection Neurons

Two types of projection neurons are present in all the species studied (see Part I, Sect. 2.1) based on morphological characteristics. Physiological data demonstrates the presence of two types of STN neurons (Beurrier et al. 1999; see Sect; 2.2.2):

1. STN neurons able to burst and generate LTS and plateau potentials
2. STN neurons that only respond with an LTS

It remains unclear whether the morphologically discerned types of STN projection neurons are comparable to the physiological subtypes. Moreover at least two types of neurons can be discerned based on the presence or absence of Ca^{2+} binding proteins (Part I, Sect. 2.3.3). There is no literature available that indicates a relation to the physiological subtypes.

6.7
Neurotransmitter Input Versus Receptors in the STN

In Part I, Sect. 2.3, several neurotransmitter receptors and modulators are described. The main neurotransmitters that are discussed in the literature are GABA, glutamate and dopamine. However, NO, here considered as a gaseous neurotransmitter, is present in the majority (95%) of the human STN neurons. Its effects are rarely studied in the STN.

The presence of dopamine in the STN is restricted (see Bevan et al. 2007). Its D1 and D2 receptors are present in the STN. These dopamine receptors are found postsynaptically in the STN (Part I, Sect. 2.3.4.1). Are these receptors, in analogy to the striatal receptors, responsible for activation and inhibition of STN neurons? It is therefore necessary to obtain the topographical distribution of these receptors, also in light of 3D modelling and the two morphological types of STN projection neurons. Analogous reasoning holds for the other receptors.

6.8
The Pedunculopontine Nucleus

Our preliminary culturing results show the enormous effect of acetylcholine on the behaviour of STN neurons. It seems clear that direct cortical information reaches the human pedunculopontine nucleus, which is responsible for the initiation of movement (Part I, Sects. 5.2.5 and 5.2.6). What function is still exerted by this nucleus during DBS stimulation of the STN? Could a false cortical signal in this nucleus be responsible for akinesia, and is high-frequency antidromic stimulation perhaps responsible for suppression of this phenomenon?

6.9
Nigro-subthalamic Connections

Our research shows that in the rat, a contralateral projection from the substantia nigra into the subthalamic nucleus exists (Part I, Sect. 6), albeit only an ipsilateral connection has been accepted until now. Because the intimate relation of substantia nigra and subthalamic nucleus, this is understandable. Studies that verify this connection in other species are missing in the literature. Nevertheless, if such a connection is present, does the dopamine and GABAergic contralateral input of the substantia nigra into the STN also not modulate the oscillatory activity of the STN? This idea is supported by unilateral dopamine lesions that decrease the neuron discharge rate in the contralateral STN, while increasing this rate in the ipsilateral STN. This last finding also supports the ideas presented in Sect. 5.7 of part I.

6.10
Another Cortico-subthalamic Loop

A short direct cortico-STN connection in humans is not supported in the literature (Part I, Sect. 5). In this monograph, arguments are brought forward for a cortical–STN loop that stays within the cerebral peduncle, leaves the peduncle not earlier than in the substantia nigra (bundle of Poppi), reaches the pedunculopontine nucleus (and colliculus superior, see Fig. 21 in Part I), and presumably the STN (Part I, Sect. 5.1). This explains perhaps the relatively rare presence of small round boutons, since the STN is presumably the end of the loop that cannot be detected easily.

6.11
Nissl-Based Subdivision of the STN

Within the STN of all the species studied in this monograph, a tri-partition is noted. The middle or central part of the STN can be recognized based on its lower content of neurons. These inherent borders can be used to show the localization of efferent and afferent connections and the somatotopy present in the STN. Moreover, it makes a comparison of the different localizations possible.

Appendix I Model Parameter Values Otsuka et al. 2004

STN parameter	STN value	STN parameter	STN value
g_{Na}	49 mS/cm^2	τ_a^o, τ_a^1	1, 1 ms
g_K	57 mS/cm^2	τ_b^o, τ_b^1	0, 200 ms
g_{IA}	5 mS/cm^2	τ_c^o, τ_c^1	45, 10 ms
g_L	15 mS/cm^2	τ_{d1}^o, τ_{d1}^1	400, 500 ms
g_T	5 mS/cm^2	τ_{d2}	130 ms
g_{Ca-K}	1 mS/cm^2	τ_m^o, τ_m^1	0.2, 3 ms
g_{leak}	0.35 mS/cm^2	τ_h^o, τ_h^1	0, 24.5 ms
v_{Na}	60 mV	τ_n^o, τ_n^1	0, 11 ms
v_K	−90 mV	τ_p^o, τ_p^1	5, 0.33 ms
v_{leak}	−60 mV	τ_q^o, τ_q^1	0, 400 ms
θ_a	−45 mV	τ_r	2 ms
θ_b	−90 mV	θ_a^r	−40 mV
θ_c	−30.6 mV	$\theta_b^{r1}, \theta_b^{r2}$	−60, −40 mV
θ_{d1}	−60 mV	$\theta_c^{r1}, \theta_c^{r2}$	−27, −50 mV
θ_{d2}	0.1	$\theta_{d1}^{r1}, \theta_{d1}^{r2}$	−40, −20 mV
θ_m	−40 mV	θ_m^r	−53 mV
θ_h	−45.5 mV	$\theta_h^{r1}, \theta_h^{r2}$	−50, −50 mV
θ_n	−41 mV	$\theta_n^{r1}, \theta_n^{r2}$	−40, −40 mV
θ_p	−56 mV	$\theta_p^{r1}, \theta_p^{r2}$	−27, −102 mV
θ_q	−85 mV	$\theta_q^{r1}, \theta_q^{r2}$	−50, −50 mV
θ_r	0.17	σ_a	−0.5
k_a	−14.7	σ_b^1, σ_b^2	−30, 10

(continued)

(continued)

STN parameter	STN value	STN parameter	STN value
k_b	7.5	σ_c^1, σ_c^2	−20, 15
k_c	−5	$\sigma_{d1}^1, \sigma_{d1}^2$	−15, 20
k_{d1}	7.5	σ_m	−0.7
k_{d2}	0.02	σ_h^1, σ_h^2	−15, 16
k_m	−8	σ_n^1, σ_n^2	−40, 50
k_h	6.4	σ_p^1, σ_p^2	−10, 15
k_n	−14	σ_q^1, σ_q^2	−15, 16
k_p	−6.7		
k_q	5.8		
k_r	−0.08		

Appendix 2 Model Parameter Values Terman et al. 2002; Rubin and Terman (2004)

STN parameter	STN value	GPe parameter	GPe value
g_{leak}	2.25 nS/μm²	g_{leak}	0.1 nS/μm²
g_K	45 nS/μm²	g_K	30 nS/μm²
g_{Na}	37.5 nS/μm²	g_{Na}	120 nS/μm²
g_T	0.5 nS/μm²	g_T	0.5 nS/μm²
g_{Ca}	0.5 nS/μm²	g_{Ca}	0.15 nS/μm²
g_{AHP}	9 nS/μm²	g_{AHP}	30 nS/μm²
v_{leak}	−60 mV	v_{leak}	−55 mV
v_K	−80 mV	v_K	−80 mV
v_{Na}	55 mV	v_{Na}	55 mV
v_{Ca}	140 mV	v_{Ca}	120 mV
τ_h^1	500 ms	τ_h^1	0.27 ms
τ_n^1	100 ms	τ_n^1	0.27 ms
τ_r^1	17.5 ms	τ_h^0	0.05 ms
τ_h^0	1 ms	τ_n^0	0.05 ms
τ_n^0	1 ms	τ_r	30 ms
τ_r^0	40 ms	ϕ_h	0.05
ϕ_h	0.75	ϕ_n	0.05
ϕ_n	0.75	ϕ_r	1
ϕ_r	0.2	k_1	30
k_1	15	k_{Ca}	20
k_{Ca}	22.5	ε	1*10⁻⁴ ms⁻¹
ε	3.75*10⁻⁵ ms⁻¹	θ_m	−38 mV
θ_m	−30 mV	θ_h	−58 mV
θ_h	−39 mV	θ_n	−50 mV
θ_n	−32 mV	θ_r	−70 mV
θ_r	−67 mV	θ_a	−57 mV
θ_a	−63 mV	θ_s	−35 mV
θ_b	0.4	θ_h^τ	−40 mV
θ_s	−39 mV	θ_n^τ	−40 mV

(continued)

(continued)

STN parameter	STN value	GPe parameter	GPe value
θ_h^τ	$-57\,\text{mV}$	a	$2\,\text{ms}^{-1}$
θ_n^τ	$-80\,\text{mV}$	σ_m	10
θ_r^τ	$68\,\text{mV}$	σ_h	-12
a	$5\,\text{ms}^{-1}$	σ_n	14
σ_m	15	σ_r	-2
σ_h	-3.1	σ_a	2
σ_n	8	σ_s	2
σ_r	-2	σ_h^τ	-12
σ_a	7.8	σ_n^τ	-12
σ_b	-0.1	σ_g^H	2
σ_s	8	θ_g^H	$-57\,\text{mV}$
σ_h^τ	-3	θ_g	$20\,\text{mV}$
σ_n^τ	-26	β	$0.08\,\text{ms}^{-1}$
σ_r^τ	-2.2	$v_{STN \to GPe}$	$0\,\text{mV}$
σ_g^H	8	$v_{GPe \to GPe}$	$-100\,\text{mV}$
θ_g^H	$-39\,\text{mV}$		
θ_g	$30\,\text{mV}$		
β	$1\,\text{ms}^{-1}$		
$v_{GPe \to STN}$	$-85\,\text{mV}$		

References

Afsharpour S (1985a) Light microscopic analysis of Golgi-impregnated rat subthalamic neurons. J Comp Neurol 236:1–13
Afsharpour S (1985b) Topographical projections of the cerebral cortex to the subthalamic nucleus. J Comp Neurol 236:14–28
Alexander GE, Crutcher MD (1990) Functional architecture of basal ganglia circuits: neural substrates of parallel processing. Trends Neurosci 13:266–271
Alexander GE, De Long MR, Strick PL (1986) Parallel organization of functionally segregated circuits linking basal ganglia and cortex. Ann Rev Neurosci 9:357–381
Bar-Gad I, Morris G, Bergman H (2003) Information processing, dimensionality reduction and reinforcement learning in the basal ganglia. Progress in Neurobiology 71:439–473
Benabid A, Benazzous A, Pollak P (2002) Mechanisms of deep brain stimulation. Mov Disord 17:S73–S74
Benabid AL (2003) Deep brain stimulation for Parkinson's disease. Curr Opin Neurobiol 13:696–706
Benazzouz A, Piallat B, Pollak P, Benabid AL (1995) Responses of substantia nigra pars reticulate and globus pallidus complex to high frequency stimulation of the subthalamic nucleus in rats: electrophysiological data. Neurosci Lett 189:77–80
Benazzouz A, Gao D, Ni Z, Benabid A-L (2000a) High frequency stimulation of the STN influences the activity of dopamine neurons in the rat. Neuroreport 11:1593–1596
Benazzouz A, Gao DM, Ni ZG, Piallat B, Bouali-Benazzouz R, Benabid AL (2000b) Effect of high-frequency stimulation of the subthalamic nucleus on the neuronal activities of the substantia nigra pars reticulate and ventrolateral nucleus of the thalamus in the rat. Neuroscience 99:289–295
Benazzouz A, Breit S, Koudsie A, Pollak P, Krack P, Benabid A (2002) Intraoperative microrecordings of the subthalamic nucleus in Parkinson's disease. Mov Disord 17:S145–S149
Ben-Paz H, Bergman H, Goldberg JA, Giladi N, Hansel D, Reches A, Simon ES (2001) Synchrony of rest tremor in multiple limbs in Parkinson's disease: evidence for multiple oscillators. J Neural Transm 108:287–296
Bergman H, Wichmann T, Karmon B, DeLong MR (1994) The primate subthalamic nucleus. II. Neuronal activity in the MPTP model of parkinsonism. J Neurophysiol 72:507–520
Beurrier C, Congar P, Bioulac B, Hammond C (1999) Subthalamic nucleus neurons switch from single spike activity to burst-firing mode. J Neurosci 19:599–609
Beurrier C, Bioulac B, Hammond C (2000) Slowly inactivating sodium current (I(NaP)) underlies single-spike activity in rat subthalamic neurons. J Neurophysiol 3:1951–1957
Beurrier C, Bioulac B, Audin J, Hammond C (2001) High-frequency stimulation produces a transient blockade of voltage-gated currents in subthalamic neurons. J Neurophysiol 85:1351–1356
Bevan MD, Wilson CJ (1999) Mechanisms underlying spontaneous oscillation and rhythmic firing in rat subthalamic neurons. J Neurosci 9:7617–7628

Bevan M, Magill P, Terman D, Bolam J, Wilson C (2002) Move to the rhythm: oscillations in the subthalamic nucleus-external globus pallidus network. Trends Neurosci 5:525

Bevan MD, Hallworth NE, Beafreton J (2007) GABAergic control of the subthalamic nucleus. In: Tepper JM, Abercrombie ED, Bolam JP (eds) GABA in the basal ganglia. Progr Brain Res 60:173-188

Boyes J, Bolam JP (2007) Localization of GABA receptors in the basal ganglia. In: Tepper JM, Abercrombie ED, Bolam JP (eds) GABA and the basal ganglia. Progr Brain Res 60:229-243

Braak H, Braak E, Yilmazer D, de Vos RA, Jansen EN, Bohl J (1996) Pattern of brain destruction in Parkinson's and Alzheimer's diseases. J Neural Transm 103:455-490

Breit S, Schulz JB, Benabid A (2004) Deep brain stimulation. Cell Tissue Res 318:275-288

Breit S, Lessmann L, Benazzouz A, Schulz JB (2005) Unilateral lesion of the pedunculopontine nucleus induces hyperactivity in the subthalamic nucleus and substantia nigra. Eur J Neurosci 2:2283-2294

Brown P (2003). Oscillatory nature of human basal ganglia activity: relationship to the pathophysiology of Parkinson's disease. Mov Disord 8:357-363

Calabresi P, Centonze D, Bernardi G (2000) Electrophysiology of dopamine in normal and denervated striatal neurons. Trends Neurosci 23:S57-S63

Dostrovsky J, Lozano A (2002) Mechanisms of deep brain stimulation. Mov Disord 7: S63-S68

Eytan D, Marom S (2006) Dynamics and effective topology underlying synchronization in networks of cortical neurons. J Neurosci 6:8465-8476

Fearnley J, Lees A (1991) Aging and Parkinson's disease: substantia nigra regional selectivity. Brain 114:2283-2301

Feger J, Hammond C, Rouzaire-Dubois B (1979) Pharmacological properties of acetylcholine-induced excitation of subthalamic nucleus neurons. Br J Pharmacolol 5:511-515

Flores G, Hernandez S, Rosales MG, Sierra A, Martines-Fong D, Flores Hernandez J (1996) M3 muscarine receptors mediate cholinergic excitation of the spontaneous activity of the subthalamic neurons in the rat. Neurosci Lett 03:203-206

Garcia L, Audin J, D'Alessandro G, Bioulac B, Hammond C (2003) Dual effect of high-frequency stimulation on subthalamic neuron activity. J Neurosci 3:8743-8751

Garcia L, D'Alessandro G, Bioulac B, Hammond C (2005a) High-frequency stimulation in Parkinson's disease: more or less? Trends Neurosci 28:209-216

Garcia L, D'Alessandro G, Fernagut P, Bioulac B, Hammond C (2005b) Impact of high-frequency stimulation parameters on the pattern of discharge of subthalamic neurons. J Neurophysiol 4:3662-3669

Gibb WRG, Lees A (1994) Pathological clues to the cause of Parkinson's disease. In: Marsden CD, Fahn S (eds) Movement disorders 3. Oxford: Butterworth Heinemann, pp 147-166

Gillies A, Willshaw D (2004) Models of the subthalamic nucleus: the importance of intranuclear connectivity. Med Eng Phys 26:723-732

Gillies A, Willshaw D (2006) Membrane channel interactions underlying rat subthalamic projection neuron rhythmic and bursting activity. J Neurophysiol 5:2352-2365

Grill W, McIntyre C (2001) Extracellular excitation of central neurons: implications for the mechanisms of deep brain stimulation. Thalamus Relat Sys 1:269-277

Gross GW (1979) Simultaneous single unit recording in vitro with a photoetched later deinsulated gold multimicroelectrode surface. IEEE Trans Biomed Eng 6:273-279

Hammond C, Yelnik J (1983) Intracellular labelling of rat subthalamic neurones with horseradish peroxidase: computer analysis of dendrites and characterization of axon arborization. Neuroscience 8:781-790

Hassler R (1938) Zur Pathologie der Paralysis agitans und des postencephalitischen Parkinsonismus. J Psychol Neurol 9:193-230

References

Heida T (2003) Electric field-induced effects on neuronal cell biology accompanying dielectrophoretic trapping. Adv Anat Embryol Cell Biol 73:1–80

Heldoorn M, Van Leeuwen JL, Vanderschoot J, Marani E (2001a) Electrotonic coupling in a network of compartmental external urethral sphincter motorneurons of Onuf's nucleus. Neurocomputing 38-40:647–658

Heldoorn M, Van Leeuwen JL, Vanderschoot J (2001b) Modelling the biomechanics and control of sphincters. J Exp Biol 204:4013–4022

Hodgkin A, Huxley AF (1952) A quantitative description of membrane current and its application to conduction and excitation in nerve. J Physiol 17:500–544

Hurtado JM, Lachaux JP, Beckley DJ, Gray CM, Sigvardt KA (2000) Inter- and intralimb oscillator coupling in parkinsonian tremor. Mov Disord 5:683–691

Izhikevich EM, Desai NS, Walcott EC, Hoppensteadt FC (2003) Bursts as a unit of neural information: selective communication via resonance. Trends Neurosci 26:1–13

Kita H, Chang HT, Kitai ST (1983) The morphology of intracellularly labeled rat subthalamic neurons: a light microscopic analysis. J Comp Neurol 15:245–257

Lozano A, Dostrovsky J, Chen R, Ashby P (2002) Deep brain stimulation for Parkinson's disease: disrupting the disruption. Lancet Neurol 1:225–231

Magnin M, Morel A, Jeanmonod D (2000) Single-unit analysis of the pallidum, thalamus and subthalamic nucleus in parkinsonian patients. Neuroscience 96:549–564

Matsumura M (2001) Experimental parkinsonism in primates. Stereotact Funct Neurosurg 7:91–97

Matsumura M, Kojima J (2001) The role of the pedunculopontine tegmental nucleus in experimental parkinsonism in primates. Stereotact Funct Neurosurg 7:108–115

McIntyre CC, Grill WM, Sherman DL, Thakor NV (2004a) Cellular effects of deep brain stimulation: model-based analysis of activation and inhibition. J Neurophysiol 1:1457–1469

McIntyre CC, Savasta M, Kerkerian-Le Goff L, Vitek JL (2004b) Uncovering the mechanism(s) of action of deep brain stimulation: activation, inhibition, or both. Clin Neurophysiol 15:1239–1248

Mink JW (1996) The basal ganglia: focused selection and inhibition of competing motor programs. Prog Neurobiol 50:381–425

Miocinovic S, Parent M, Butson CR, Hahn PJ, Russo GS, Vitek JL, McIntyre CC (2006) Computational analysis of subthalamic nucleus and lenticular fasciculus activation during therapeutic deep brain stimulation. J Neurophysiol 6:1569–1580

Mogilner AY, Benabid AL, Rezai AR (2001) Brain stimulation: current applications and future prospects. Thalamus Relat Sys 1:255–267

Montgomery E, Gale J (2005) Mechanisms of deep brain stimulation: implications for physiology, pathophysiology and future therapies: 10th Annual Conference of the International FES Society

Montgomery EB, Baker KB (2000) Mechanisms of deep brain stimulation and future technical developments. Neurol Res 2:259–266

Nakanishi H, Kita H, Kitai ST (1987) Electrical membrane properties of rat subthalamic neurons in an in vitro slice preparation. Brain Res 37:35–44

Nakanishi H, Kita H, Kitai ST (1988) An N-methyl-D-aspartate receptor mediated excitatory postsynaptic potential evoked in subthalamic neurons in an in vitro slice preparation in the rat. Neurosci Lett 5:130–136

Nambu A (2005) A new approach to understand the pathophysiology of Parkinson's disease. J Neurol 52, suppl :IV/1–IV/4

Nambu A, Tokuno H, Hamada I, Kita H, Imanishi M, Akazawa T, Ikeuchi Y, Hasegawa N (2000) Excitatory cortical inputs to pallidal neurons via the subthalamic nucleus in the monkey. J Neurophysiol 4:289–300

Nambu A, Tokuno H, Takada M (2002) Functional significance of the cortico-subthalamo-pallidal 'hyperdirect' pathway. Neurosci Res 3:111–117

Otsuka T, Abe T, Tsukagawa T, Song WJ (2000) Single compartment model of the voltage-dependent generation of a plateau potential in subthalamic neurons. Neurosci Res Suppl 4:581

Otsuka T, Murakami F, Song WJ (2001) Excitatory postsynaptic potentials trigger a plateau potential in rat subthalamic neurons at hyperpolarized states. J Neurophysiol 6:1816–1825

Otsuka T, Abe T, Tsukagawa T, Song W-J (2004) Conductance-based model of the voltage-dependent generation of a plateau potential in subthalamic neurons. J Neurophysiol 2:255–264

Pascual A, Modolo J, Beuter A (2006) Is a computational model useful to understand the effect of deep brain stimulation in Parkinson's disease? J Integr Neurosci 5:541–559

Plenz D, Kitai ST (1999) A basal ganglia pacemaker formed by the subthalamic nucleus and external globus pallidus. Nature 400:677–682

Raz A, Vaadia E, Bergman H (2000) Firing patterns and correlations of spontaneous discharge of pallidal neurons in the normal and tremulous 1-methyl-4-phenyl-1,2,3,6 tetrahydropyridine vervet model of parkinsonism. J Neurosci 0:8559–8571

Romijn HJ, van Huizen F, Wolters PS (1984) Towards an improved serum free, chemically defined medium for long-term culturing of cerebral cortex tissue. Neurosci Biobeh Rev 8:301–334

Romo R, Schultz W (1992) Role of primate basal ganglia and frontal cortex in the internal generation of movements. III. Neuronal activity in the supplementary motor area. Exp Brain Res 1:396–407

Rosales MG, Flores G, Hernandez S, Martinez-Aceves J (1994) Activation of subthalamic neurons produces NMDA receptor-mediated dendritic dopamine release in substantia nigra pars reticulate: a microdialysis study in the rat. Brain Res 45:335–337

Rubin JE, Terman D (2004) High-frequency stimulation of the subthalamic nucleus eliminates pathological thalamic rhythmicity in a computer model. J Comp Neurosci 16:211–235

Rudy B, McBain CJ (2001) Kv3 channels: voltage-gated K+ channels designed for high-frequency repetitive firing. Trends Neurosci 24:517–526

Rutten W, Mouveroux J-M, Buitenweg J, Heida C, Ruardij T, Marani E, Lakke E (2001) Neuroelectronic interfacing with cultured microelectrode arrays toward a cultured probe. Proc IEEE 89:1013–1029

Segev R, Baruchi I, Hulata E, Ben Jacob E (2004) Hidden neuronal correlations in cultured networks. Phys Rev Lett 2:118–102

Shen KZ, Johnson SW (2000) Presynaptic dopamine D2 and muscarine M3 receptors inhibit excitatory and inhibitory transmission to rat subthalamic neurons in vitro. J Physiol 25:331–341

Shen M, Piser TM, Seybold VS, Thayer S (1996) Cannabinoid receptor agonists inhibit glutamatergic synaptic transmission in rat hippocampus. J Neurosci 6:4322–4334

Song W-J, Baba Y, Otsuka T, Murakami F (2000) Characterization of Ca2+ channels in rat subthalamic nucleus neurons. J Neurophysiol 4:2630–2637

Squire LR, Bloom FE, McConnell SK, Roberts JL, Spitzer NC, Zigmond MJ (2003) The Basal Ganglia: Fundamental neuroscience, 2nd edn. Academic Press, New York, pp 815–839

Stegenga J, le Feber J, Marani E, Rutten WLC (2007) Analysis of cultured neuronal networks using intra-burst firing characteristics. IEEE Trans Biomed Eng (in press)

Suri RE, Albani C, Glattfelder AH (1997) A dynamic model of motor basal ganglia functions. Biol Cybern 6:451–458

References

Tam DC (2002) An alternate burst analysis for detecting intra-burst firings based on inter-burst periods. Neurocomputing 44–46:1155–1159

Tang J, Moro E, Lozano A, Lang A, Hutchison W, Mahant N, Dostrovsky J (2005) Firing rates of pallidal neurons are similar in Huntington's and Parkinson's disease patients. Exp Brain Res 166:230–236

Terman D, Rubin JE, Yew AC, Wilson CJ (2002) Activity patterns in a model for subthalamopallidal network of the basal ganglia. J Neurosci 2:2963–2976

Thomas CA Jr, Springer PA, Loeb GA, Berwald-Netter Y, Okun LM (1972) A miniature electrode array to monitor the bioelectrical activity of cultured cells. Exp Cell Res 4:61–66

Usunoff KG, Itzev DE, Ovtscharoff WA, Marani E (2002) Neuromelanin in the human brain: a review and atlas of pigmented cells in the substantia nigra. Arch Physiol Biochem 10:257–369

van Dorp R, Jalink K, Oudega M, Marani E, Ypey DL, Ravesloot JH (1990) Morphological and functional properties of rat dorsal root ganglion cells cultured in a chemically defined medium. Eur J Morphol 8:430–444

van Elburg RAJ, van Ooyen A (2004) A new measure for bursting. Neurocomputing 58–60:497–502

van Pelt JPS, Wolters, Corner MA et al. (2004) Long-term characterization of firing dynamics of spontaneous bursts in cultured neural networks. IEEE Trans Biomed Eng 1:2051–2062

van Welsum RA, Van der Voet GB, Marani E, Van Keep J-P, De Wolff FA (1989) Aluminum affects interconnections between aggregates of cultured hippocampal neurons. J Neurol Sci 3:157–166

Vitek J (2002) Mechanisms of deep brain stimulation: excitation or inhibition. Mov Disord 17:S69–S72

Wagenaar DA, DeMarse TB et al. (2005a) MeaBench: A toolset for multi-electrode data acquisition and online analysis. International IEEE EMBS Conference on Neural Engineering, Arlington, VA

Wagenaar DA, Madhavan R, Pine J, Potter SM (2005b) Controlling bursting in cortical cultures with closed-loop multi-electrode stimulation. J Neurosci 5:680–688

Wichmann T, DeLong MR (1996) Functional and pathophysiological models of the basal ganglia. Curr Opin Neurobiol 6:751–758

Wigmore MA, Lacey MG (2000) A Kv3-like persistent, outwardly rectifying, Cs^+-permeable, K^+ current in rat subthalamic nucleus neurones. J Physiol 27:493–506

Wilson CL, Puntis M, Lacey MG (2004) Overwhelmingly asynchronous firing of rat subthalamic nucleus neurons in brain slices provides little evidence for intrinsic interconnectivity. Neuroscience 123:187–200

Wu Y, Levy R, Ashby P, Tasker R, Dostrovsky J (2001) Does Stimulation of the GPi control dyskinesia by activating inhibitory axons? Mov Disord 6:208–216

Index

Acetylcholine 24, 33, 35–37, 77
Action potential 12, 15, 43, 53, 64
   after hyperpolarization 15, 16, 27, 55
Activity pattern
   burst firing 17–18, 45, 65
   single spike 15–17
   spontaneous activity 15–18, 33–35, 38, 43–45, 53–55, 65
Axons 11–13, 27, 70, 75, 76

Basal Ganglia (BG)
   centre-surround model 5
   direct pathway 2, 4–7
   hyperdirect pathway 4–6
   indirect pathway 2–5, 7, 67
   role 1, 2, 4, 5, 9
   thalamocortical circuit 2, 6
Bursts 27–31
   burst detection 28–30
   mixed burst mode 18
   network bursts 28–31
   pure burst mode 17
   rebound bursts 56–58, 60, 65, 67, 72

Cortico-subthalamic connection 4, 78

Deep brain stimulation (DBS) 1, 9, 60, 72
   mechanisms 11–14, 26, 62, 73
Dopamine 3, 6, 7, 24, 27, 31, 40, 77
   D1 receptors 6, 7
   D2 receptors 6, 7
Dynorphin 70

Electrical stimulation 9, 11, 13, 37–39
Enkephalin 2, 70

GABA 2, 13, 31, 75, 77
Globus Pallidus
   externus (GPe) 2, 7, 8, 67
   GPe neuron model 67

GPi neuron model 71, 72
   internus (GPi) 2, 4, 7, 8, 10, 27
Glutamate 13, 24, 27, 31, 77

High frequency stimulation 11, 12, 14, 26–27, 38
Hodgkin and Huxley 40
Input currents
   depolarizing 19–22, 25, 26, 45, 46, 55, 58, 72
   hyperpolarizing 19–22, 45, 46, 56, 59, 68
Interneurons 11, 66, 75, 76
In-vitro recordings
   brain slices 14–31
   dissociated cell cultures 14, 31–39

Low-threshold spike (LTS) 21–24
   membrane potential 21, 23

Membrane dynamics 41, 52, 63
   L-type Ca^{2+} channels 22, 24, 26, 48, 52, 58, 65
   N-type Ca^{2+} channels 24, 33
   T-type Ca^{2+} channels 24, 26, 48, 58, 59, 65
Membrane input resistance 23, 25
Multi electrode array (MEA) 1, 28, 31–34, 37

Negative slope conductance 15, 16, 44, 51, 55, 59
Network architectures 35, 68–71
Network model
   GPe-STN 67–68, 71
   GPe-STN-GPi 71–72
   inter-nuclear 67–74
   intra-nuclear 66

Parkinson's disease 1, 6–9
   MPTP-induced parkinsonism 1, 8, 70
   neuronal firing pattern 9
   neuronal firing rate 8–9
   parkinsonian 5, 14, 40, 60, 61, 67, 72

Pedunculopontine nucleus (PPN)   37, 76–78
Plateau potential   21–23, 26, 45, 56, 66
  ionic mechanisms   23–25
  membrane potential   21, 22, 24, 25
  non-plateau-generating neurons   23, 25, 60
  plateau-generating neurons   23, 25, 40, 60
  stability index   22
Projection neurons   8, 11, 63, 76–77

Striatum   2–4, 6–9, 12, 31, 68, 70, 71, 73
Substance P   2

Substantia nigra
  pars compacta (SNc)   2, 8
  pars reticulate (SNr)   2, 27
Subthalamic nucleus (STN) models   8–10
  multi-compartment model   63–66
  single-compartment model   40, 41, 52, 60, 67
Synaptic inputs   24–26, 28, 40, 68

Thalamus   2–7, 9, 10, 13, 25, 70, 72, 73
  thalamic neuron model   71

Printing: Krips bv, Meppel, The Netherlands
Binding: Stürtz, Würzburg, Germany